my **revisi⏻n** notes

WJEC GCSE

DESIGN AND TECHNOLOGY

Ian Fawcett
Jacqui Howells
Andy Knight
Chris Walker

HODDER
EDUCATION
AN HACHETTE UK COMPANY

Although every effort has been made to ensure that website addresses are correct at time of going to press, Hodder Education cannot be held responsible for the content of any website mentioned in this book. It is sometimes possible to find a relocated web page by typing in the address of the home page for a website in the URL window of your browser.

Hachette UK's policy is to use papers that are natural, renewable and recyclable products and made from wood grown in well-managed forests and other controlled sources. The logging and manufacturing processes are expected to conform to the environmental regulations of the country of origin.

Orders: please contact Hachette UK Distribution, Hely Hutchinson Centre, Milton Road, Didcot, Oxfordshire, OX11 7HH. Telephone: +44 (0)1235 827827. Email education@hachette.co.uk Lines are open from 9 a.m. to 5 p.m., Monday to Friday. You can also order through our website: www.hoddereducation.co.uk

ISBN: 978 1 5104 7170 2

© Ian Fawcett, Jacqui Howells, Andy Knight and Chris Walker 2019

First published in 2019 by

Hodder Education,

An Hachette UK Company

Carmelite House

50 Victoria Embankment

London EC4Y 0DZ

www.hoddereducation.co.uk

Impression number 10 9 8 7 6 5 4

Year   2023

Cover photo © Shining Black – stock.adobe.com

Typeset in India.

Printed and bound by CPI Group (UK) Ltd, Croydon, CR0 4YY

A catalogue record for this title is available from the British Library.

Get the most from this book

Everyone has to decide his or her own revision strategy, but it is essential to review your work, learn it and test your understanding. These Revision Notes will help you to do that in a planned way, topic by topic. Use this book as the cornerstone of your revision and don't hesitate to write in it: personalise your notes and check your progress by ticking off each section as you revise.

Tick to track your progress

Use the revision planner on pages 4–5 to plan your revision, topic by topic. Tick each box when you have:

- revised and understood a topic
- tested yourself
- practised the exam questions and checked your answers.

You can also keep track of your revision by ticking off each topic heading in the book. You may find it helpful to add your own notes as you work through each topic.

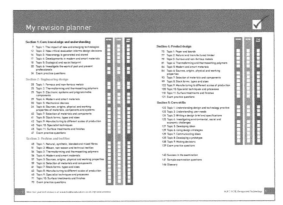

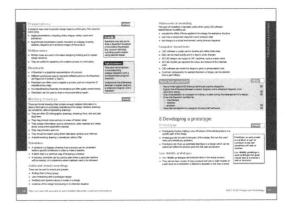

Features to help you succeed

Exam tips

Expert tips are given throughout the book to help you polish your exam technique in order to maximise your chances in the exam.

Typical mistakes

The authors identify the typical mistakes students make and explain how you can avoid them.

Key words

Clear, concise definitions of essential key terms are provided where they first appear and in the Glossary at the back of the book.

Now test yourself

Short, knowledge-based questions provide the first step in testing your learning. Answers can be found online at www.hoddereducation.co.uk/myrevisionnotes.

Exam practice

Practice exam questions are provided for each topic. Use them to consolidate your revision and practise your exam skills.

My revision planner

Countdown to my exams

6–8 weeks to go

- Start by looking at the specification — make sure you know exactly what material you need to revise and the style of the examination. Use the revision planner on pages 4 and 5 to familiarise yourself with the topics.
- Organise your notes, making sure you have covered everything on the specification. The revision planner will help you to group your notes into topics.
- Work out a realistic revision plan that will allow you time for relaxation. Set aside days and times for all the subjects that you need to study, and stick to your timetable.
- Set yourself sensible targets. Break your revision down into focused sessions of around 40 minutes, divided by breaks. These Revision Notes organise the basic facts into short, memorable sections to make revising easier.

REVISED ☐

2–6 weeks to go

- Read through the relevant sections of this book and refer to the exam tips, exam summaries, typical mistakes and key terms. Tick off the topics as you feel confident about them. Highlight those topics you find difficult and look at them again in detail.
- Test your understanding of each topic by working through the 'Now test yourself' questions in the book. Look up the answers online at **www.hoddereducation.co.uk/ myrevisionnotes**
- Make a note of any problem areas as you revise, and ask your teacher to go over these in class.
- Look at past papers. They are one of the best ways to revise and practise your exam skills. Write or prepare planned answers to the exam practice questions provided in this book. Check your answers online and try out the extra quick quizzes at **www.hoddereducation. co.uk/myrevisionnotes**
- Use the revision activities to try out different revision methods. For example, you can make notes using mind maps, spider diagrams or flash cards.
- Track your progress using the revision planner and give yourself a reward when you have achieved your target.

REVISED ☐

One week to go

- Try to fit in at least one more timed practice of an entire past paper and seek feedback from your teacher, comparing your work closely with the mark scheme.
- Check the revision planner to make sure you haven't missed out any topics. Brush up on any areas of difficulty by talking them over with a friend or getting help from your teacher.
- Attend any revision classes put on by your teacher. Remember, he or she is an expert at preparing people for examinations.

REVISED ☐

The day before the examination

- Flick through these Revision Notes for useful reminders, for example the exam tips, exam summaries, typical mistakes and key terms.
- Check the time and place of your examination.
- Make sure you have everything you need — extra pens and pencils, tissues, a watch, bottled water, sweets.
- Allow some time to relax and have an early night to ensure you are fresh and alert for the examinations.

REVISED ☐

My exams

Unit 1: Design and Technology in the 21st Century

Date:..

Time:...

Location:...

1 Core knowledge and understanding

1 The impact of new and emerging technologies

Examples of how new and emerging technologies have changed industry and enterprise are described below.

- Following the Industrial Revolution the use of steam to provide power led to the development of innovative machinery and manufacturing equipment which allowed products to be produced more quickly and efficiently.
- The use of electricity to power large items of machinery led to products being **mass produced** on **assembly lines**.
- Modern factories increasingly use **automated production**. Robots are used to complete some repetitive and monotonous tasks previously carried out by humans. Productivity and product quality are improved as a result of **automation**.

Figure 1.1 **Robots are used in the production of printed circuit boards**

> **Mass produced**: hundreds or thousands of identical products manufactured on a production line.
>
> **Assembly line**: a line of equipment/machinery manned by workers who gradually assemble a product as it passes along the line.
>
> **Automated production**: the use of computer-controlled equipment or machinery in manufacturing.
>
> **Automation**: the use of automatic equipment in manufacturing.

Market pull and technology push

REVISED

Some products are developed as a result of market research, which uncovers a need for a new product or a demand from consumers/users for an improved existing product. This is referred to as **market pull**.

Technological developments in materials, components or manufacturing methods lead to the development of new or improved products. These new products are a result of **technology push**. For example:

- many touchscreen devices exist only because of the development of the material graphene
- the technology now exists to seamlessly weave **conductive** threads into fabric for clothing, which will interact directly with the wearer.

> **Market pull**: a new product is developed in response to a demand in the market or users.
>
> **Technology push**: products developed as a result of new technology.
>
> **Conductive**: the ability to transmit heat or electricity.

Consumer choice

- Designers and manufacturers respond to customer choice by developing products that specifically meet the needs of consumers.
- Many people are influenced by the latest technology and will only want to buy products that include it. Many products such as mobile phones are replaced because products with the latest technology become available.

Product life cycle

A product's **life cycle** is a marketing strategy that looks at the four main stages a product goes though from its introduction to the market through to its decline in terms of sales.

Table 1.1 **The four main stages of a product's life cycle**

Introduction	Following an advertising campaign, the new product is introduced to the marketplace
Growth	Sales will grow as consumers become aware of the product and buy it
Maturity	Sales are at their peak, with companies hoping to achieve maximum sales for the product
Decline	Sales begin to fall as most interested consumers now own the product or a new product has replaced it

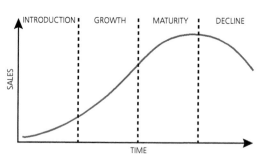

Figure 1.2 **The product life cycle**

The length of a product life cycle will depend on the product. For example, classic fashion styles maintain good sales for many years, while fashion **fad products** become popular very quickly with rapid growth in sales but decline just as quickly as new styles are introduced. Examples of fad products include loom bands and fidget spinners.

> **Fad product:** a product that is highly popular for only a very limited amount of time.

People, culture and society

REVISED

Global production and its effects on people and culture

- Developments in transport makes it easier for manufacturers to ship materials, components and products all over the world.
- The global economy allows for materials and components to be sourced in one country and products or part products to be manufactured in another and then shipped around the world.
- Manufacturing costs can be reduced through automation.
- Developments in mobile technology and the internet make it easier for us to communicate with people all over the world. This leads to greater competition among manufacturers, which in turn keeps prices low.

There are, however, downsides to this global society which directly affect workers:

- Designers need to be aware and sensitive to cultural differences between societies. What is acceptable in one country may be offensive in another. The values, beliefs and customs of different cultures should be respected.
- Global production is a threat to traditional industries, skills and techniques in some developing countries:
  - Importing cheap products from overseas and not buying locally-produced products can lead to job losses in our own society.
  - The use of automation in manufacturing leads to job losses. Fewer people are needed in factories.
  - Workers overseas are often paid low wages in an effort to keep costs down and maximise profit for the manufacturer.
  - The use of mobile technology can make people feel isolated as the opportunities for face-to-face interaction are limited.

Now test yourself answers at **www.hoddereducation.co.uk/myrevisionnotes**

Legislation

- The law states that all products sold to consumers must be safe and that responsibility for safety lies with manufacturers, producers, distributors and retailers.
- Failing to comply with legal requirements can lead to fines or even imprisonment.
- Producers must warn consumers of any potential risks attached to the use of a product.

British Standards Institute (BSI)

- The BSI sets safety standards and specifications for a range of products.
- Products are thoroughly tested and inspected and if they meet the required standards they are awarded the BSI kitemark to show that they are safe and of a high quality.

International Organization for Standardization (ISO)

- The ISO develops and publishes international standards for materials, products, processes and services, focusing on shared challenges and things that are important to consumers.
- A group of international experts and stakeholders discusses what the standard should be and when agreement is reached the standard is published.
- Each standard is given an individual ISO number. There are more than 22,000 international standards.

Consumer rights

- The Consumer Rights Act 2015 protects consumers when goods purchased or services provided are not as expected.
- The Act also covers digital products or buying online, and covers contracts such as those issued with mobile phones.
- The law states that all goods should be as described or seen when purchased and be fit for purpose.
- The Act protects consumers against faulty or **counterfeit** goods and poor service or problems with builders; this includes rogue traders.
- Consumers can request a refund, repair or replacement when goods purchased do not meet certain standards. Products should:
  - function as intended
  - be of satisfactory quality
  - be as described at the time of purchase.
- If a service provided fails to meet expectations and a full refund or replacement is not possible, then the provider is legally bound to offer some form of **compensation**.

> **Counterfeit**: an imitation of something, sold with the intent to defraud.
>
> **Compensation**: payment given to someone as a result of loss.

The Trades Descriptions Act 1988

- This Act makes it illegal to apply a false trade description to goods and services. This covers:
  - the materials used to make a product
  - the size of the product

   ○ the product's fitness for purpose

   ○ performance characteristics including information relating to testing products.

- The Act is enforced by local Trading Standards authorities and criminal charges can be brought for false claims about a product or service.

Moral and ethical factors

- A global market allows unrestricted trade. Many people are able to grow their businesses, make a profit and improve the lives of their workers by offering regular employment and an income.
- In a global economy, not everyone is treated fairly. There is no obligation on companies to improve the lives of their workers. Some companies put profit above all else, with low wages and poor working conditions.
- Some companies follow a more ethical approach to trade. They focus on goods and services that are beneficial to consumers, they show that they are socially responsible by treating their workers fairly with acceptable rates of pay and good working conditions, and they support environmental causes.
- Ethical traders are open and transparent about costs. It is important to them that trade is seen to be fair.
- Some companies choose not to disclose their costs because this could reveal poor wages or working conditions if profits are revealed and considered high. This is particularly true in the textile industry.

Sustainability and the environment

REVISED

- Designers, manufacturers and consumers are increasingly aware of the negative impact that new technologies and the development and disposal of products have on the environment.
- **Sustainability** is about meeting today's needs without compromising the needs of future generations.
- It is important to look at ways of reducing the environmental impact.
- In hybrid cars that use both diesel or petrol and an electric motor, fuel consumption is reduced, along with the levels of carbon dioxide (CO_2) emitted. This is a significant improvement for the environment. Many people believe fully electric cars are the way forward.
- Developments in renewable energy are allowing us to make better use of alternative sources of energy and to reduce our reliance on **finite fossil fuels** such as coal or oil.
- Many polymers are difficult to recycle. New technology is being developed that will allow these polymers to be broken down more effectively and safely, enabling them to be recycled.

> **Sustainability:** meeting today's needs without compromising the needs of future generations.
>
> **Finite fossil fuels:** a limited amount of resources that cannot be replaced.

Computer-aided design (CAD)

- Technological developments in CAD packages have changed the way designers work. All aspects of developing design ideas through to 3D models can be done on the computer.
- CAD packages allow for changes to be made or errors to be rectified.
- CAD models allow designers and manufacturers to simulate how products will look and perform in different situations.

- Emerging **cloud-based technology** allows for collaborative work – designers can share projects via the internet in the cloud. This collaboration can be on a global scale and therefore cuts the need for travel.
- **Generative design** is a new development that makes use of mathematical **algorithms** based on set parameters or design requirements.
- Disadvantages of CAD include high initial set-up costs, such as training employees, and the possibility of losing work through computer failure or a virus.

Cloud-based technology: technology that allows designers to share content via the internet.

Generative design: a computer-based iterative design process that generates a number of possibilities that meet certain constraints, including potential designs that would not previously have been thought of.

Algorithm: a logical computer-based procedure for solving a problem.

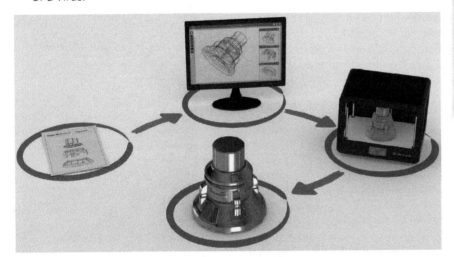

Figure 1.3 **Stages in the development of a product – initial sketch, CAD drawing, 3D printed prototype and the final product**

Computer-aided manufacture (CAM)

- CAM machinery can manufacture products and components directly from CAD drawings.
- In industry, CAM machines are often used where high volumes of identical products of a consistent high quality are needed.
- Initial set-up costs for these machines can be high but they are considered more efficient in the long term as they can run for long periods without breaks.
- Disadvantages of CAM include high set-up costs, the impact on the workforce with loss of employment, technological failure and on-going maintenance costs.

Using CAM equipment

Table 1.2 **Types of CAM equipment**

CNC embroidery machine	Designs can be embroidered directly onto a range of textile fabrics
	Designs can be saved then repeated several times with the same high-quality finish
Vinyl cutters	A pattern based on a CAD drawing can be cut from a roll of self-adhesive vinyl
	Letters used in signage are often cut on a vinyl cutter; the colour is determined by the vinyl

Figure 1.4 **3D printing or additive manufacture**

CNC router	A rotating router cutter follows a CAD drawing to cut a path or shape
	Alternative cutting tools can alter the profile of the cut
	The CAD drawing dictates the depth of the cut
Laser cutter	Laser cutters use a laser beam to cut through (or vaporise) material; material can also be engraved
	Intricate patterns can be cut from a variety of materials. However, not all materials can be cut as some, such as nylon and PVC, burn or melt
3D printer	3D printing is also known as **additive manufacture**. It uses a thermoforming polymer roll or spool of filament which is heated, then extruded through a head to form a layer. The bed then moves down for the next layer to be printed
	The strength of the product is determined by the inner design of the print and the material used

> **Additive manufacture:** computer-controlled manufacture of a 3D object by adding together materials layer by layer.

Figure 1.5 **An intricate laser-cut design on a designer dress**

Typical mistake

When a question asks for an advantage of CAD or CAM, make sure you fully explain your answer. It is not enough to state CAD and CAM are quicker, easier or faster unless they are compared to an alternative method. For example, if a designer wishes to present ideas in different colours, CAD allows this easily, whereas a hand-drawn sketch has to be drawn several times, which is time consuming.

2 How critical evaluation informs design decisions

Sustainability and environmental issues when designing and making

REVISED

It is important that designers take the environment into consideration when making design decisions. For example:

- choosing materials that are more environmentally friendly
- manufacturing products using efficient, low-energy processes
- ensuring better build quality in products so that they last longer
- using less or no packaging, or using recycled packaging
- reducing transportation by using local manufacturers with locally sourced materials
- using LED lights instead of filament lamps
- designing products to last and avoiding those with a short life cycle
- considering what happens to products once they are no longer needed; making recycling of products easier
- considering Fairtrade products, where everyone in the supply chain is treated fairly.

Environmental directives (laws), which come from the European Union or organisations such as the World Energy Council, are a set of targets for governments in all countries to work towards in an effort to reduce energy consumption, reduce pollution and eliminate the disposal of hazardous waste into the environment. These directives also cover climate change, air pollution and the protection of wildlife.

Social, cultural, economic and environmental responsibilities

- Designers and manufacturers need to consider the views of consumers – demand for more environmentally friendly products is increasing.
- In an effort to reduce energy consumption, domestic appliances carry an energy rating label like the one shown in Figure 1.6. A+++ is the most efficient while G is the least. The most efficient products will also help reduce household energy bills.

Figure 1.6 Energy rating on a washing machine

Linear and circular economy

Table 1.3 **Linear and circular economy**

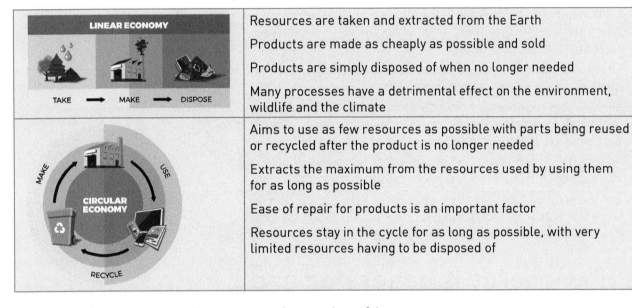

	Resources are taken and extracted from the Earth
LINEAR ECONOMY	Products are made as cheaply as possible and sold
	Products are simply disposed of when no longer needed
	Many processes have a detrimental effect on the environment, wildlife and the climate
CIRCULAR ECONOMY	Aims to use as few resources as possible with parts being reused or recycled after the product is no longer needed
	Extracts the maximum from the resources used by using them for as long as possible
	Ease of repair for products is an important factor
	Resources stay in the cycle for as long as possible, with very limited resources having to be disposed of

- Increasingly, consumers are choosing not to buy products if they are not environmentally friendly.
- Manufacturers are having to reconsider the way they source, manufacture and package products.
- Laws are in place to tackle pollution and the ways in which waste is disposed of.
- Cradle-to-cradle production links with the circular economy – from the source of materials to a product's rebirth as a new product, there is little or no waste.
- Cradle-to-grave production includes consideration of how the product will ultimately be disposed of.

Life-cycle analysis (LCA)

- A life-cycle analysis of a product looks at the environmental impact of a product throughout its entire life, from the source of materials through its useful life to final disposal and potential rebirth as a new product.
- In a life-cycle analysis, the following factors should be considered: source of raw materials, materials processing, manufacturing, use, end of life and transportation, including energy used at various points during its life cycle.

> **Typical mistake**
>
> Do not confuse life-cycle analysis (LCA) with a product life cycle. Life-cycle analysis looks at the environmental impact of a product over its entire lifetime, whereas the product life cycle is a marketing strategy looking at sales of products.

Design obsolescence

- Some products are designed or manufactured in a way that limits their life cycle. This is called planned obsolescence.
- For example, a newer model of a mobile phone might be designed with a different connector so that the old charging cable can no longer be used.
- The advantage of obsolescence to a designer and manufacturer is that demand will continue for new products, even from customers who already own the product, as they will need to update their obsolete model with a newer version.
- The disadvantage is that designers must find new ways to keep ahead of the competition, which will require research and being able to predict trends.

Carbon footprint

- Carbon footprint is a measure of the total amount of greenhouse gases produced as a result of human activity, which includes the manufacturing of products.
- Greenhouse gases are usually measured in units of carbon dioxide and are said to be the cause of global warming.
- Every time we use energy from fossil fuels, we add to our carbon footprint – for example, heating our homes or workplaces with gas, oil or coal emits CO_2 into the atmosphere.
- Transportation of products or travel by car or aeroplane uses energy derived from fossil fuels, adding to our carbon footprint.
- Designers can reduce the carbon footprint of a product by adopting more sustainable approaches to design, for example sourcing locally-produced materials, which cuts down the transportation of raw materials.

3 How energy is generated and stored

Energy

REVISED

Energy is needed to:
- manufacture products and power products and systems
- cause something to move, heat something and create light and sound
- process materials: extract, mould, bend, cut, drill, print and join materials.

Table 1.4 **Types of renewable and non-renewable energy sources**

Renewable energy sources	
Source	**Explanation**
Wind	A wind turbine extracts energy from the wind. The blades are connected to a generator which produces electricity
Solar	Photovoltaic (PV) panels produce electricity when exposed to sunlight
Geothermal	Cold water is pumped underground and heated by the Earth's heat. It can be used to heat homes or used in power stations and converted to electricity
Hydroelectric	Dams, which house large turbines, are built to trap water. When the water is released, the pressure turns the turbines, generating electricity

Renewable energy sources	
Wood/biomass	Wood not used in the timber industry is chipped and used as fuel instead of burning coal. This can provide heat for homes or used to generate electricity
	In some biomass schemes, plants such as soy are grown to produce materials which can be processed into biofuels
Wave	It is possible to harness energy from waves on the sea, although it is not widely used. In future, tidal power offers the possibility of extracting energy from the rise and fall of the tide

Non-renewable energy sources	
Source	**Explanation**
Coal	Coal is mined from the ground and burned in power stations to generate electricity
Oil	Crude oil is extracted from the Earth and refined into liquid fuels such as petrol. It can also be used to generate electricity in power stations
Gas	Gas is extracted through drilling and piped through the national grid to houses and factories. It can also be used to generate electricity in power stations
Nuclear	Uranium ore is mined from the Earth and transformed into nuclear fuel. This is used in a nuclear generator to generate heat and then converted to electricity

Issues surrounding the use of fossil fuels

REVISED

- Waste such as CO_2 and pollutants such as sulphur dioxide are emitted into the atmosphere when fossil fuels are burned. This can cause breathing problems and also contributes to global warming.
- Fossil fuels cannot be replaced and will eventually run out.
- Fossil fuels have high energy density – they hold a lot of chemical energy per kilogram of fuel, making them ideal for transportation. Batteries, such as those currently used in electric cars, are heavy, offer a limited range and take too long to charge. Currently they cannot compete with the ease of using petrol.

The advantages and disadvantages of renewable energy sources

- Renewable sources of energy are non-polluting and considered to be better for the environment.
- Although biomass fuels release CO_2 as they are burned, trees are replanted which absorb CO_2 as they grow. The process is classed as **carbon neutral**.
- Initial outlay for the equipment needed for extracting renewable energy is expensive. However, following installation, it produces free energy.
- Wind and solar power depend on weather conditions and therefore cannot be relied on. Some people find wind farms and solar panels unsightly.
- In order to build dams used to generate hydroelectric power, valleys in rural areas are flooded. This can damage the natural habitat for wildlife.
- Geothermal energy units are expensive and reliant on the underground rocks being hot near the surface.
- Increasingly, manufacturers are investing in renewable energy to power their factories and are installing equipment to recover waste energy from various processes to heat their offices. This will reduce energy bills and demonstrates a more ethical approach to manufacturing.

> **Carbon neutral:** no net release of carbon dioxide into the atmosphere – carbon is offset.

Renewable energy sources for products

Compact renewable energy sources can be used in some products:

- Small solar PV panels can produce a small current to recharge a battery. Flexible solar PV panels that can charge a mobile phone can be found on clothing and bags.
- Low-powered products can be charged from a small wind generator.
- Electronic road signs are often powered by a solar PV panel.
- Clockwork wind-up mechanisms can provide a temporary source of power for mechanical or electronic products.

Energy generation and storage in a range of contexts

Motor vehicles

- Electric cars use batteries as their energy source; they are recharged by being plugged into an electricity source.
- Electric cars do not produce any emissions, although the power to charge them comes from fossil fuels.
- Batteries take hours to fully recharge and the car's range is limited.
- Electric cars are efficient; some kinetic energy can also be recovered when the driver uses the brakes. This energy can then be stored in the battery.
- Electric cars are increasing in popularity as they are cheap to run and more environmentally friendly.
- Rechargeable hybrid cars offer a greater driving range but also lower emissions.

Mains-powered products

- Many products such as household appliances are charged using mains electricity.
- Products left on standby are a concern as they continue to use electricity even when they are not being used.

Battery-powered products

- Energy is stored in the rechargeable batteries of many products, such as mobile phones, tablets and cordless products.
- Solar panels absorb energy from the sun during daylight hours, passing the power to a battery as an electrical charge. This can be used to power products such as garden lights.
- Some products such as torches or television remote controls are powered by non-renewable batteries, which need to be replaced.
- As batteries contains chemicals, they should be disposed of sensibly through appropriate recycling schemes.

Figure 1.7 Electronic road sign powered by solar energy

Typical mistake

'Analyse' or 'Evaluate' questions require depth of knowledge and understanding, with sound reasoning or judgements made within the answer. You will lose marks if your answers are not sufficiently detailed or are merely descriptive, or if you offer no reasoning or judgements.

Exam tip

Make sure you fully understand the meaning of key words and terminology. This will help you to understand the context of each question and enable you to write a full and detailed answer.

Now test yourself

TESTED ☐

1 Explain how automation used in industry is changing the way products are manufactured. [4]
2 Describe a situation where a consumer would be entitled to compensation from a service provider. [2]
3 Biomass fuels are said to be carbon neutral. Explain the meaning of this term. [2]
4 Describe in detail one advantage and one disadvantage of wind power. 2 × [2]
5 Explain the benefits to the environment of the 'circular economy'. [3]

4 Developments in modern and smart materials

Smart materials

REVISED

Smart materials change or react to a change in their environment such as temperature, light, pressure or electrical input. Reactions include a change in colour, shape or resistance.

Shape memory alloys (SMA)

- Shape memory alloys return to their original shape if heated.
- Possible uses include medical applications such as medical fastenings used in bone fractures.

Polymorph

- Polymorph is a thermoforming polymer supplied in granular form. When heated in water to 62°C it softens and forms a pliable volume of material that can be moulded and shaped.
- It solidifies on cooling and can be modelled and shaped with hand tools or machinery.
- If reheated in water it becomes pliable once again.
- Polymorph is a useful material for model making and prototyping, and is ideally suited to school projects.

1 Granules of polymorph   2 Add hot water   3 Lift out of water when soft

4 Mould to shape

Figure 1.8 The four stages of polymorph

Photochromic pigment

- Photochromic pigments or dyes change colour in response to changes in light. For example, sunglasses can change colour in response to UV radiation.

Thermochromic pigment

- Thermochromic pigments or dyes change colour in response to a change in heat and can be engineered to specific heat ranges.
- Thermochromic dyes can be used in baby bottles to give an indication of the temperature of the milk.

Figure 1.9 A thermochromic mug changes colour when boiling water is poured into the mug

Micro-encapsulation

- **Micro-encapsulation** is a process of applying microscopic capsules to fibres, fabrics, paper and card.
- The capsules can contain vitamins, therapeutic oils, moisturisers, antiseptics and anti-bacterial chemicals which are released through friction.

Biomimetics

- **Biomimicry** is when the inspiration for new materials, structures and systems comes from the natural world.
- Fastskin®, developed by Speedo, mimics the shark's natural sandpaper-like skin by reducing drag in the water. It is used for performance-enhancing swimwear.

> **Micro-encapsulation:** tiny microscopic droplets containing various substances applied to fibres, yarns and materials, including paper and card.
>
> **Biomimicry:** taking ideas from and mimicking nature.

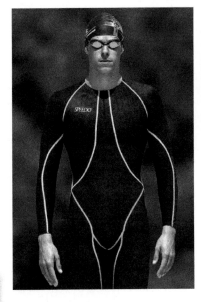

Figure 1.10 The Fastskin swimsuit is an example of biomimicry

5 Ecological and social footprint

Changing society's view on waste

- Designers are gradually being forced to produce products which have a minimal impact on the Earth, due to increasing consumer awareness, through the television and internet, of the problems caused by a **throwaway society**.
- There is increasing focus on the use of polymers in products and packaging. There has been an effort to reduce the amount of polymers used, for example by bringing in charges for the use of plastic carrier bags. Wales was one of the first countries to introduce these charges.
- Electronic and mechanical products may contain hundreds of different components from a wide range of raw materials, including toxic materials such as lead, cadmium, mercury sulphuric acid and radioactive substances.

> **Throwaway society:** a society that excessively consumes and wastes resources.

If products are disposed of incorrectly:

- they can end up in a **landfill site**
- hazardous materials can leak out of them into the environment, where they can get into the water system and cause serious health problems.

The Waste Electrical and Electronic Equipment (WEEE) directive helps to reduce the damage caused by waste products by:

- making manufacturers and producers take responsibility for what happens to their products at the end of the products' lives
- requiring retailers to offer a free take-back service for old products, which must then be disposed of at an approved facility
- making a requirement for local councils to provide recycling facilities for electronic products.

Other issues related to recycling electronic products include:

- realising that some materials that can be recovered are valuable, such as copper and gold
- designing the product to be easily separated into component materials
- accepting that redesigning products to be more recyclable can raise costs initially.

Consumer pressure and market competition are likely to force designers and manufacturers to produce more environmentally friendly products.

Ecological footprint

- An **ecological footprint** is a measure of the impact human activity has on the environment – in simple terms, the demand made by people on the world's natural resources.
- We depend on the world's natural resources and productive land to enable us to produce the goods and services which sustain and support most modern lifestyles.
- The products and clothes we use daily, the food we eat, the waste we generate and the way we live all contribute to our ecological footprint.
- Biologically-productive land is continually being cleared to sustain our lifestyles and an increasing global population.
- Humanity's ecological footprint is currently the equivalent of 1.7 Earths.
- If we continue to use up the world's natural resources more quickly than nature can replace them, we will create an **ecological deficit**.

The Six Rs of sustainability

The types of questions a designer might ask that could reduce the environmental impact of a product are outlined in Table 1.5.

Table 1.5 Questions a designer might ask when considering sustainability

Rethink	Is there a better way of making the product that is less harmful to the environment? Can the design be simplified to make manufacturing easier?
Recycle	Can the product be recycled easily after it is no longer needed? Can the components and materials be separated easily? Can recycled materials be used?
Repair	Can this product be repaired easily if it breaks? Can the component parts be replaced easily?
Refuse	Consumers might not buy the product if it is not environmentally friendly. Where will it be made and what are the conditions for workers? Is the product unethical?
Reduce	Can the number of component parts or new materials be reduced? Can packaging be reduced? Can the manufacturing process be simplified to reduce the energy used?
Reuse	Can any parts be reused after they are no longer needed? Could the product be reused for something else once its primary use is no longer required?

Living in a greener world

- Protecting the environment by reducing the amount of greenhouse gases we produce and by reducing pollution and waste in general requires a co-ordinated approach from all countries.
- Consumers can have an impact by modifying their behaviour, for example by using energy more efficiently, buying products from producers who are committed to making products with green credentials, and recycling household waste.
- Government directives can also have an effect – for example, the introduction of charges for plastic carrier bags reduced the number of bags used by 83 per cent.
- Product designers need to find ways to make products more efficient. For example, new developments in lighting such as the development of more efficient LED lighting and the use of sensors and timers to control lights in schools and offices lead to savings in the use of energy.
- Smart heating systems that can be controlled from a smartphone and adapt to the user's lifestyle are another example of how new developments can save resources.

Fairtrade

- Fairtrade sets up partnership schemes between producers, businesses and consumers that offer a better approach for all involved. Social, economic and environmental standards are set for all companies, producers and workers involved in the supply chain.
- By setting up fairer trading conditions, with workers having a share of the profits or fairer wages, the lives of workers will be improved and it will help combat poverty.
- Conditions for workers must be of a satisfactory standard, no one must be **exploited** and workers' rights must be protected.
- Consumers can be satisfied that when they buy products that carry the Fairtrade mark, they are positively supporting disadvantaged workers and producers in developing countries.

6 Investigate the work of past and present professionals

Engineering design

REVISED

Apple

- Much of Apple's success is down to the importance the company places on innovation and design.
- Apple was a pioneer in the use of graphical user interfaces (GUIs).
- In 1983, the Apple Lisa computer was the first desktop computer to have icons or small images to represent files, folders and discs and a cursor controlled by a mouse. The cursor and mouse system is still used today despite the introduction of touchscreen technology.
- Apple products are easily distinguishable by their sleek design and consistent shapes, colours and materials.
- British industrial designer Jonathan Ives was responsible for the styling of the first iPod and iMac. Aesthetics and user experience were put at the forefront of Apple's design philosophy.
- Apple is often criticised for its products being developed with planned **obsolescence**. The company is also criticised for software updates that do not work on older products and for developing its own ports for connecting other devices.

> **Obsolescence:** when a product is out of date or no longer usable.

James Dyson

See 'Product design' below.

Shigeru Miyamoto

- Shigeru Miyamoto initially trained as an industrial designer, but after starting work at Nintendo® he created some of the best known video game characters, including Mario and Luigi and Donkey Kong.
- He wanted to create a more immersive experience in which users had greater choice over the path their character took, leading to games such as *The Legend of Zelda* series.
- Miyamoto was also concerned with the way in which the user interacted with the gaming environment and was responsible for developments in controllers such as shoulder buttons for the NES controller, the thumb-operated joystick on the N64 controller and the modular handset technology on the Nintendo Switch™.
- Miyamoto's approach to design is based on his experience of playing the game and getting feedback from a wide audience rather than just experienced gamers.
- In recent years Miyamoto has been less involved with individual games and has taken a role overseeing all areas of Nintendo's development and product range.

Laura Ashley

- Laura Ashley was founded by Merthyr-born Laura Ashley and her husband Bernard.
- Laura initially designed napkins, table mats and tea towels. She wanted to make patchwork quilts but was unable to find suitable printed fabric, so Bernard designed a printing process that enabled her to print her own designs.
- The brand was established through selling headscarves, which Laura had seen girls wearing in Italy and which were popularised when the actress Audrey Hepburn wore them in the film *Roman Holiday*.
- The business grew rapidly, with the first shop in Machynlleth and the first factory in Carno in Montgomeryshire becoming the centre of a multi-national business.
- Laura started designing dresses for social wear instead of work towards the end of the 1960s, and her long Victorian-inspired silhouette proved very popular, becoming known as the 'Laura Ashley look'.
- The business expanded into home furnishings and affordable fabrics, allowing customers to co-ordinate their homes.
- Part of the company ethos was to use natural materials such as pure cotton for clothing and home furnishings, recycled paper for wallpaper and rainforest-friendly timber for wooden furniture.

Stella McCartney

- Stella McCartney is a designer with a global reputation for stylish design.
- After an apprenticeship with Savile Row tailor Edward Sexton and graduation from Central Saint Martins, McCartney opened her own shop in London selling silk slip dresses. These become her trademark design.
- She was appointed as a designer at Parisian fashion house Chloé, which hoped she would attract a younger clientele. She was soon appointed creative director and in 2000 was awarded VH1/Vogue Fashion Awards' designer of the year. McCartney then set up her own business.
- Her signature style features tailored garments with very feminine lingerie-inspired clothing. It is important to McCartney that women feel comfortable in her clothes and that they are easy to wear.
- She has a green consciousness, refusing to use leather or fur in her collections and using alternatives such as:
  - recycled polyester for her Falabella bag
  - re-engineered cashmere
  - silk from suppliers that do not exploit silk worms
  - renewable energy to power her shops and providing recycled shopping bags.

Orla Kiely

- Orla Kiely is a designer with a wide range of influences, from the green Formica cupboards and orange ceilings in the kitchen of her childhood home to yellow gorse and wild flowers.
- After collecting degrees in both Print Design and Graphics and Knitwear Design, she exhibited a range of hats made from colourful wool fibres needle-punched into gingham that was bought by Harrods.
- In 2000, Kiely created her iconic print design Stem. It has simple graphic strength and charm. The printed Stem bags established her signature style – clean, simple, measured and bold.
- The design was developed into different colourways, printed on to laminated cotton fabric and used for bags, starting a trend for products that are durable and can be wiped clean.
- Orla Kiely embraced technology at an early stage, recognising the advantages of being able to manipulate, edit, recolour and resize her designs. However, she prefers to sketch her initial ideas before transferring them to computer.

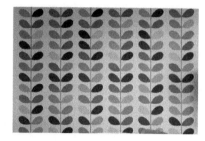

Figure 1.11 **A version of Orla Kiely's Stem design**

Product design

Airbus

- Airbus is a European company best known for aircraft but which also has divisions that focus on helicopters, military equipment and space travel.
- The Airbus A380 is the world's largest passenger aircraft, carrying up to 800 passengers.
- Engineers overcame issues around the size and weight of the A380 aircraft by:
  - using lightweight composite materials to reduce overall weight
  - looking at biomimicry for inspiration, in particular the shape and structure of eagle wings to overcome issues with the wing span. They installed wing tips on the A380's wings; without these the aircraft's wing span would be too big for most airports.
- **Generative design** is used to optimise components or parts to reduce weight but maintain strength.
- Airbus uses CAD technology throughout development, including 3D printing of parts from a variety of materials, particularly titanium, because of its excellent strength-to-weight ratio.
- Airbus is a global company: wings are manufactured in Britain, the rear fuselage in France and the front fuselage in Germany.
- The Airbus Beluga is the company's super-transporter as transportation of parts is vital to the company's success. The huge capacity of the Beluga reduces the environmental impact as fewer journeys are needed to transport parts.

James Dyson

- James Dyson is known for creating innovative products that use new technology and engineering principles and improve on existing products.
- One of his early products was the ball barrow, a variation on the traditional wheelbarrow. The ball barrow spread the load, making it easier to push on soft ground, and improved manoeuvrability.
- Previously, vacuum cleaners relied on bags to house dust that had been collected; Dyson noted the more full the bags became, the more the suction levels dropped, making the vacuum cleaner less effective. The DC01 vacuum cleaner used cyclonic technology to collect dust without the need for a bag.
- Dyson went through 5127 iterations of the DC01 product over ten years before gaining market interest, which lead to eventual success.
- Other products developed under the Dyson name include washing machines, fans, hand dryers and hairdryers. All go through rigorous and extensive testing before the products are launched.

Figure 1.12 **Dyson ball barrow**

Bethan Gray

- Bethan Gray is a Welsh furniture designer who has designed for companies including Habitat, John Lewis and Harrods, as well as running her own studio manufacturing high-end furniture.
- She trained in 3D Design before her degree work was spotted by the head of design at Habitat, resulting in her becoming the company's design director. At Habitat she gained a wealth of experience in furniture design and retail.
- Gray then set up her own company, becoming known for using luxurious materials such as brass, marble, solid wood and leather, and for the use of highly skilled, traditional craft skills and modern manufacturing techniques.
- Many of her influences come from the forms and architectural shapes she has encountered on her worldwide travels. Her appreciation of Islamic form and aesthetics led her to become a founding partner of the Ruby Tree Collection, which aims to maintain traditional Islamic craft skills and use of materials.

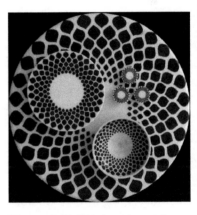

Figure 1.13 **Petals: pieces from the Ruby Tree Collection**

> **Exam tip**
>
> Read exam questions carefully to make sure you fully understand what the question is asking before attempting the answer. Underline or highlight key words to help you focus on what is important.

> **Typical mistake**
>
> Avoid writing about personal or biographical details relating to the designers you have studied. This will not gain any marks in an exam. Questions will test what you know and understand about the designers' work – their design thinking, the inspiration behind their work and any influence they may have on product design and manufacturing.

Now test yourself

TESTED

1 Explain how micro-encapsulation when used in medical dressings is beneficial to a patient. [2]
2 Discuss the use of polymorph as a modelling medium for product designers. [4]
3 Explain how biomimicry inspired engineers when developing the Airbus A380. [2]
4 Describe the features that make Apple products easily recognisable. [3]
5 Describe the impact Dyson's DC01 vacuum cleaner has had on the design of other vacuum cleaners. [3]
6 Explain how Miyamoto's work has had a lasting impact on the gaming world many players enjoy today. [4]
7 Fabric choice is important to Stella McCartney's design ethic. Describe three fabrics that have been developed as a direct result of her 'green values'. 3 × [2]
8 Describe Orla Kiely's style of work and explain how her childhood influenced her design ethos. [4]
9 Explain what is meant by the 'Laura Ashley look'. [3]
10 List three key features of Bethan Gray's work. [3]

Exam practice

1 (a) New products are designed and developed as a result of market pull and technology push. Explain how market pull and technology push have impacted on the design and development of fitness trackers. [4]
  (b) Describe how studying a product's life cycle benefits a manufacturer. [3]
2 Sustainability and environmental issues are important considerations when developing new products.
  (a) Define the term 'sustainable design'. [2]
  (b) Explain how the circular economy benefits the environment. [4]
3 Smart materials can change their properties or appearance in response to external stimuli.
  (a) State what makes shape memory alloys (SMAs) different from other metals. [1]
  (b) Describe an example where thermochromic pigment used in a named product could benefit the user. [3]
4 Electronic devices and circuits can be embedded into textile fabrics for clothing.
  (a) Heart rate monitors can be embedded into fabric. Describe how this would benefit a user with a heart condition. [2]
  (b) Micro-encapsulation is often used in medical textiles. Give two examples of its use in medicine and explain the benefit to the patient. [2 × 3]
5 Testing and modelling are essential strategies in the iterative process of designing and developing new products. Explain how Dyson exemplifies this process when designing and developing new products. [5]

ONLINE

2 Engineering design

1 Ferrous and non-ferrous metals

See Section 4, Topic 3 for details of the classification and properties of ferrous and non-ferrous metals. See Section 4, Topic 6 for details of their source, origins and working properties.

2 Thermoforming and thermosetting polymers

See Section 4, Topic 4 for details of the working properties of thermoforming and thermosetting polymers. See Section 4, Topic 6 for details of their physical and mechanical properties.

3 Electronic systems and programmable components

- Electronic **systems** are used to provide functionality to products and processes.
- Electronic systems can be broken down into **subsystems**, which can be classified as inputs, processes or outputs.
- A system diagram (sometimes called a block diagram) shows how the subsystems are connected and how the signals flow between them.
- Signals can be digital or analogue.

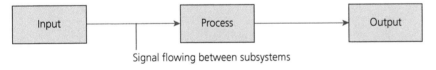

Signal flowing between subsystems

Figure 2.1 **A generic electronic system**

> **System:** a set of parts which work together to provide functionality to a product.
>
> **Subsystem:** the interconnected parts of a system.
>
> **Microcontroller:** a miniaturised computer, programmed to perform a specific task and embedded in a product.
>
> **Integrated circuit (IC):** a miniaturised, highly complex circuit in a single component.

Process subsystems

REVISED

- The process subsystem receives signals from the inputs and responds in a specific way to control the outputs. The way it responds depends on the needs of the product.
- Process subsystems can be made from semiconductor devices such as a **microcontroller**, microprocessor or computer.
- A microcontroller is a miniaturised computer that is programmed to perform a specific task.
- A microcontroller is an example of an **integrated circuit (IC)**. An IC is a miniaturised, highly complex circuit in a single component.

Table 2.1 **Process devices**

Process devices can perform functions such as:	Example application
Counting	Sports scoreboard, digital clock, pedometer
Switching	Night light, electric kettle, automatic door
Timing	Security light, burglar alarm, cooking timer

Inputs

- Inputs consist of sensors, which can monitor and measure a range of physical quantities.
- A sensor produces an electrical signal which can be digital or analogue.
- A **digital sensor** detects a yes/no situation, such as 'is the button pressed?'.
- A switch is a digital sensor, and many types of switches are available for use in products.
- An **analogue sensor** can measure how big a quantity is, such as 'how bright is the light?' or 'what is the temperature?'.

Light-dependent resistor (LDR)

- An **LDR** is an analogue sensor used to sense light.
- Its resistance falls as the light level increases.
- LDRs are used in street lamps, night lights, digital clocks (to control display brightness), CCTV cameras (to switch to night vision mode), etc.

> **Digital sensor**: a sensor to detect a yes/no or an on/off situation.
>
> **Analogue sensor**: a sensor to measure how big a physical quantity is.
>
> **LDR**: light-dependent resistor. An analogue component to sense light level.
>
> **Thermistor**: an analogue component to sense temperature.

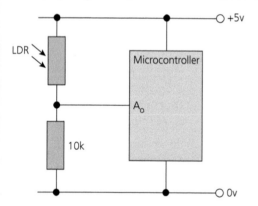

Figure 2.2 **An LDR connected to an analogue input of a microcontroller**

Thermistor

- A **thermistor** is an analogue sensor used to sense temperature.
- Its resistance falls as temperature increases.
- Thermistors are used in ovens, room thermostats, electric heaters, car engines, etc.

Outputs

- An output subsystem converts an electrical signal into a desired function.
- A buzzer produces a sound output. Buzzers are useful for providing feedback that a user has pressed a button. They are found in many products, including burglar alarms, microwave ovens, dishwashers, kitchen timers, etc.

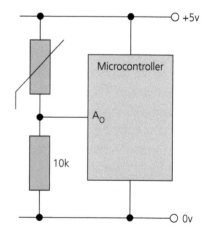

Figure 2.3 **A thermistor providing an input signal to a microcontroller**

- A light-emitting diode (LED) produces a light output.
- LEDs are available in a range of colours, sizes and shapes.
- LEDs can be used as an indicator, for example as a 'power on' light.
- LEDs can be used as a source of illumination, for example in a torch.
- A resistor must be used with an LED to limit the current flowing, or the LED will burn out.

Feedback in control systems

- **Feedback** in a system is when a signal from the output is taken and fed back into the input of the process subsystem.
- Feedback allows a microcontroller to monitor the effect of the changes it makes to its output devices.
- Feedback allows a system to achieve precise control.
- An example of feedback is in an electric oven control system.

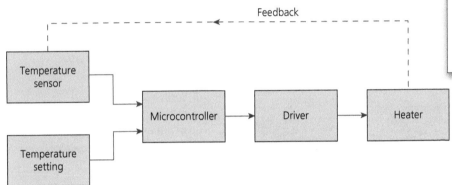

Figure 2.4 **Using an LED as an output from a microcontroller**

> **Feedback:** achieving precise control by feeding information from an output back into the input of a control system.

Figure 2.5 An oven control system, where a signal from the output is fed back to the input

Programmable systems

Microcontrollers

- A programmed microcontroller is embedded into a product to provide functionality and to enhance the performance of the product.
- Microcontrollers can be quickly reprogrammed, which is useful during product development or for product upgrades.
- Microcontrollers are found in many products, including toasters, TVs, microwave ovens, hi-fi systems, cars, etc.
- Microcontrollers can be interfaced with a wide range of digital and analogue input and output devices.
- Some output devices require a driver to boost the output signal from the microcontroller.
- A **programmable interface controller (PIC)** is a popular type of microcontroller IC. PICs are used in many GCSE projects.

Flowchart programs

- A microcontroller **program** is a set of instructions which tells the microcontroller what to do.
- When a program is run, the microcontroller executes the instructions at extremely high speed.

> **Programmable interface controller (PIC):** A microcontroller IC used in many products.
>
> **Program:** a set of instructions which tells the microcontroller what to do.

- A **flowchart** is a graphical way of showing a program.
- Standard symbols are used in flowchart programs.

Examples of flowchart commands include:

- **Input/Output** commands: 'Read the temperature sensor', 'Turn on LED', 'Turn off buzzer', etc.
- **Process** commands: 'Wait 2 seconds', 'Add 1 to the value of variable A', etc.
- **Decision** commands: 'Has temperature dropped below 5°C?', 'Is the button pressed?', 'Is A>45?'.

Subroutines

- **Subroutines** (also called 'macros') can be used to help simplify the structure of a complex program.
- A subroutine is a set of program instructions that performs a specific task, e.g. 'flash an LED five times' or 'measure how long a user holds a button pressed'.
- Subroutines are called from the main program using a **Call** command.
- A **Return** command at the end of the subroutine returns the flow back to the main program.

Symbol	Name
	Start/end
→	Arrows
	Input/Output
	Process
	Decision

Figure 2.6 **Flowchart symbols**

> **Flowchart:** a graphical representation of a program.
>
> **Subroutine:** a small subprogram within a larger program.

Exam tip

Questions may ask you to identify suitable input/output devices for a particular application, or to classify devices into digital and analogue types. You may be asked to complete a system diagram for a given application, so practise drawing system diagrams for several familiar products, including systems with feedback.

Questions on programmable devices may require you to draw (or to complete) a flowchart program, so learn the flowchart symbols and practise drawing flowcharts for the control of familiar products.

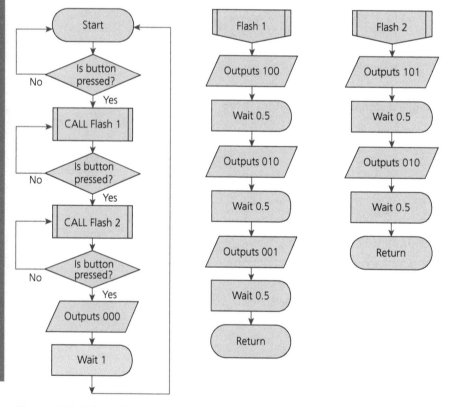

Figure 2.7 **A flowchart program using two subroutines**

Typical mistakes

Confusing digital and analogue sensors and signals is a common error. Make sure you know which sensor belongs to which category.

Other errors include incorrectly describing how the resistance of LDRs and thermistors change with light and temperature, forgetting to use a resistor with an LED, and using the wrong flowchart symbols for input/output, process and decision commands.

Now test yourself

1 Name and draw the circuit symbols for two input sensing components. [4]
2 Describe the difference between a digital and an analogue input signal. [2]
3 Explain what is meant by 'embedding a microcontroller in a product'. [3]
4 Draw a system diagram for a bicycle theft alarm, using a microcontroller as the process subsystem. Explain how the sensor(s) you have chosen works in this application. [6]
5 Draw a flowchart program, using standard symbols, for a security light which turns on for 30 seconds if movement is detected. [4]

4 Modern and smart materials

Smart materials

Electroluminescent material

- **Electroluminescent** wire (EL wire) is made from a thin copper wire core that is coated in phosphor powder. It produces a glowing light when exposed to an alternating electric **current**.
- Electroluminescent technology is also found in flexible films or thin panels. The light-emitting phosphor is sandwiched between a pair of conductive electrodes and subjected to an AC current to create light. The brightness of the light depends on the **voltage** applied.
- EL films are slowly replacing traditional LCD displays because they are flexible, do not generate heat, and are more reliable and durable.

Quantum tunnelling composite (QTC)

- **Quantum tunnelling composites** are flexible polymers that contain conductive nickel particles that can be either a conductor of electricity or an insulator.
- The nickel particles make contact with each other and are compressed when force is applied, leading to an increase in conductivity. When the force is removed, the material returns to its original state and becomes an electrical insulator.

Conductive polymers

- Polymers are generally resistant to electrical conduction, making them ideal for use in casings for electrical products.
- Conductive polymers are being developed which act as an electrical conductor. This means they can be used to replace glass and metal in products that include electronic components such as LEDs, OLED TV screens and smart windows.
- Conductive polymers are cheaper, lightweight and flexible, which means they can be used for applications that would previously have been challenging to produce.

> **Electroluminescent:** materials that provide light when exposed to a current.
>
> **Current:** a measure of the actual electricity flowing, in amps (A).
>
> **Voltage:** the electrical 'pressure' at a point in a circuit, in volts (V).

Figure 2.8 The Groclock uses electroluminescent technology

> **Quantum tunnelling composites:** materials that can change from conductors to insulators when under pressure.

5 Mechanical devices

Types of motion

There are four types of **motion**:

- **Rotary motion**: movement in a circular path, e.g. wheels, electric motor.
- **Linear motion**: movement in a straight line, e.g. car, conveyor belt.
- **Oscillating motion**: movement back and forth in a circular path, e.g. electric toothbrush head, pendulum.
- **Reciprocating motion**: movement back and forth in a straight line, e.g. needle on a sewing machine, jigsaw blade.

Rotary motion:

$$\text{Rotational speed} = \frac{\text{number of revolutions}}{\text{time taken}}$$

Linear motion:

$$\text{Speed} = \frac{\text{distance travelled}}{\text{time taken}}$$

Oscillating and reciprocating motion:

$$\textbf{Frequency} \text{ (oscillation speed)} = \frac{\text{number of oscillations}}{\text{time taken}}$$

> **Motion:** when an object moves its position over time.
>
> **Rotary motion:** movement in a circular path.
>
> **Linear motion:** movement in a straight line.
>
> **Oscillating motion:** movement back and forth in a circular path.
>
> **Reciprocating motion:** movement back and forth in a straight line.
>
> **Frequency:** the number of pulses produced per second, in hertz (Hz).

- Rotational speed is often measured in units of 'revolutions per minute (rpm)' or, sometimes, 'revolutions per second' (rps). To convert revolutions per second into rpm, multiply by 60.
- Speed is measured in various units, so it is necessary to look carefully at the units given in an exam question. Typical units are metres per second (ms^{-1}), kilometres per hour (km h^{-1}) or millimetres per second (mm s^{-1}).
- To convert mm s^{-1} into ms^{-1}, divide by 1000.

Mechanical systems

- Mechanical systems can produce different types of movement.
- Mechanical systems can change the magnitude (size) and direction of forces and movement.
- A mechanical system will take an input **force** (or motion) and process it to produce an output force (or motion).
- A force is a push, a pull or a twist.
- A simple **mechanism** trades off forces against distances moved. If one increases, the other must decrease.

> **Mechanism:** a series of parts that work together to control forces and motion.
>
> **Force:** a push, a pull or a twist.

Mechanical components

Levers

- A **lever** is a **rigid** bar that pivots on a **fulcrum**. The input force is called the **effort** and the output force is called the **load**.
- The lever arm length is the distance between the force and the fulcrum.
- If the input arm length is larger than the output arm length, the lever will increase the force applied but reduce the distance moved by the force. In other words, the load will be larger than the effort.

> **Rigid:** inflexible, stiff.
>
> **Lever:** a rigid bar that pivots on a fulcrum.
>
> **Fulcrum:** the pivot point on a lever.
>
> **Effort:** the input force on a lever.
>
> **Load:** the output force from a lever.

- The lever arm length is in inverse proportion to the force. If the output arm is half as long as the input arm, the output force will be twice as big as the input force.

$$\frac{\text{Load}}{\text{Effort}} = \frac{\text{input arm length}}{\text{output arm length}}$$

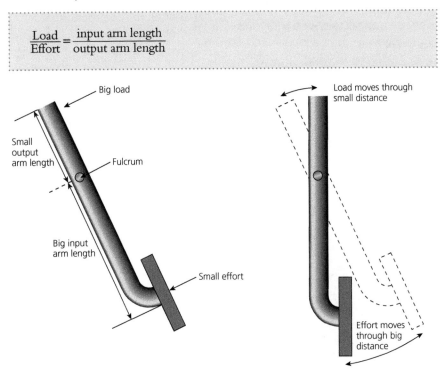

Figure 2.9 **A brake pedal is a lever**

Linkages

- A **linkage** is used to direct forces and movement to where they are needed.
- A simple pulley changes the direction of motion of a cord.
- A reverse motion linkage will reverse the direction of input motion.
- A bell crank will transfer motion around a corner.
- A peg and slot converts rotary motion into oscillating motion.
- A crank and slider converts rotary motion into reciprocating motion.

Linkage: a component to direct forces and movement to where they are needed.

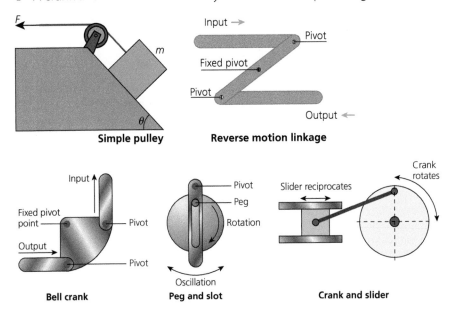

Figure 2.10 **Examples of linkages**

Cams

- A **cam** and follower converts rotary motion into reciprocating motion.
- A snail cam causes the follower to rise steadily, followed by a sudden drop.
- A pear-shaped cam creates a sudden rise and fall followed by a long period where the follower does not move.
- The eccentric cam creates an even rise and fall motion throughout its rotation.
- Cams are used in toys, machinery and engines.

> **Cam:** a component used with a follower to convert rotary motion to reciprocating motion.

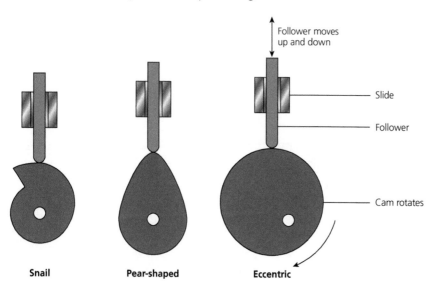

Snail   Pear-shaped   Eccentric

Figure 2.11 **Three types of cam**

Gears

- Gear systems transfer rotary motion.
- A **simple gear train** consists of two **spur gears**, which are wheels that interlock with teeth around their edge.
- The input gear is called the **driver gear**. The output gear is called the **driven gear**.
- If the driver is the smaller gear, it is sometimes called a **pinion**.
- The two gears will rotate in opposite directions.
- The smaller gear will rotate faster than the larger gear.
- The number of teeth on the gear is in inverse proportion to the speed it rotates. A gear with twice as many teeth will rotate at half the speed.
- Gear systems are used in cordless drills, clocks, winches and cars.

> **Simple gear train:** two spur gears meshed together.
>
> **Spur gear:** a gear wheel with teeth around its edge.
>
> **Driver gear:** the input gear on a gear train.
>
> **Driven gear:** the output gear from a gear train.

$$\frac{\text{Input gear speed}}{\text{Output gear speed}} = \frac{\text{number of teeth on driven gear}}{\text{number of teeth on driver gear}}$$

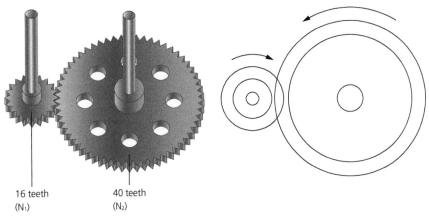

16 teeth
(N₁)

40 teeth
(N₂)

Figure 2.12 **A simple gear train**

Belt drives

- A **pulley and belt** drive transfers rotary motion, like a gear system.
- The input and output pulleys can be separated by a large distance by using a long belt.
- The smaller pulley will rotate faster than the larger pulley.
- The input and output pulleys rotate in the same direction.
- Belt drives are used in washing machines, workshop bench drills and car engines.

Rack and pinion

- A rack and pinion converts rotary motion into linear motion.
- Examples of uses of rack and pinion mechanisms are stairlifts and sliding doors.

Figure 2.13 **A belt drive in a washing machine**

Figure 2.14 **A rack and pinion**

> **Typical mistake**
>
> A common error is confusing the directions of motion in lever systems and simple gear trains, and confusing whether these systems increase or decrease the magnitude of forces and of movement.

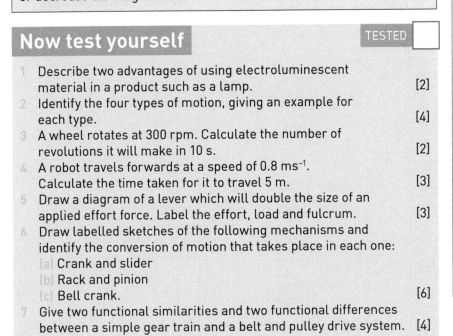

Now test yourself
TESTED

1 Describe two advantages of using electroluminescent material in a product such as a lamp. [2]
2 Identify the four types of motion, giving an example for each type. [4]
3 A wheel rotates at 300 rpm. Calculate the number of revolutions it will make in 10 s. [2]
4 A robot travels forwards at a speed of 0.8 ms⁻¹. Calculate the time taken for it to travel 5 m. [3]
5 Draw a diagram of a lever which will double the size of an applied effort force. Label the effort, load and fulcrum. [3]
6 Draw labelled sketches of the following mechanisms and identify the conversion of motion that takes place in each one:
   (a) Crank and slider
   (b) Rack and pinion
   (c) Bell crank. [6]
7 Give two functional similarities and two functional differences between a simple gear train and a belt and pulley drive system. [4]

> **Exam tip**
>
> Questions may ask you to sketch examples of mechanisms, using arrows to indicate movement, so practise sketching a variety of mechanisms and mechanical components.
>
> Make sure you can use, and rearrange, the speed – distance – time formula.
>
> Questions may test your knowledge and understanding of how linkages and mechanisms change the type of motion, and how they change the magnitude and direction of forces, so make sure you understand this for the various mechanisms.

6 Sources, origins, physical and working properties of materials, components and systems

Operational amplifiers

Voltage amplifier

- An **amplifier** turns a small signal (V_{in}) into a larger signal (V_{out}).
- The amplification factor is called the **voltage gain**.

$$\text{Voltage gain} = \frac{V\,out}{V\,in}$$

- A voltage amplifier can be made using an **op-amp**.
- Amplifiers are used in audio entertainment systems.

> The gain of the voltage amplifier circuit is controlled by two resistors.
>
> $$\text{Voltage gain} = 1 + \frac{Rf}{Ra}$$

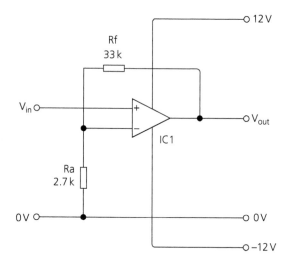

Figure 2.15 **Voltage amplifier using an op-amp IC**

Logic gates

- Logic gates are digital components.
- They process signals which are either logic 1 (high/on) or logic 0 (low/off).
- The output of a logic gate depends on the logic state of its inputs.
- You are required to know about six different gates: NOT, AND, OR, NAND, NOR and EOR.
- Each gate has a truth table which explains how the gate behaves. You need to learn the truth tables and the logic gate symbols shown in Figure 2.16.

> **Amplifier:** a subsystem to increase the amplitude of an analogue signal.
>
> **Voltage gain:** the amplification factor of an amplifier subsystem.
>
> **Operational amplifier (op-amp):** a specialised IC amplifier component.

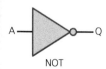

NOT

A	Q
0	1
1	0

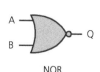

NAND

A	B	Q
0	0	1
0	1	1
1	0	1
1	1	0

AND

A	B	Q
0	0	0
0	1	0
1	0	0
1	1	1

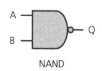

NOR

A	B	Q
0	0	1
0	1	0
1	0	0
1	1	0

OR

A	B	Q
0	0	0
0	1	1
1	0	1
1	1	1

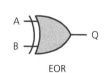

EOR

A	B	Q
0	0	0
0	1	1
1	0	1
1	1	0

Figure 2.16 **Logic gates and truth tables**

Output components

Light output

- LEDs are available in a variety of sizes, shapes and colours.
- Standard-brightness LEDs are used as indicators, e.g. an 'on' light.
- High-brightness LEDs are used in place of light bulbs for room illumination.
- LEDs have a positive terminal (the anode) and a negative terminal (the cathode).
- The cathode lead is usually the shorter lead, and it is often marked with a flat section on the case.
- A resistor must be placed in series with an LED to limit the current flowing.

The resistor value can be calculated using a modified version of the Ohm's law formula:

$$R = \frac{V_S - V_{LED}}{I}$$

where V_S is the power supply voltage, V_{LED} is the voltage drop across the LED and I is the current flowing through the LED.

Sound output

- A buzzer produces a tone when it receives power.
- A siren is a particularly loud or noticeable buzzer.
- A loudspeaker is used to produce a music or speech type of output sound. It cannot produce sound unless it receives a sound waveform.
- A piezo-sounder is a miniature loudspeaker, often used to produce simple 'beep' tones.

Movement output

- Electric motors are components which produce rotary motion.
- A MOSFET (transistor) **driver** is needed to allow a process subsystem to control a motor.
- A **back emf** diode is needed when using a MOSFET to drive a motor (or a solenoid or relay). The diode protects the MOSFET by removing the back emf generated by the motor.

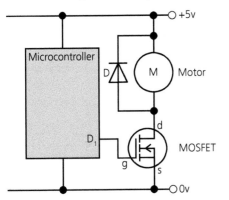

Figure 2.18 **A MOSFET and back emf diode used to drive a motor**

Figure 2.17 **The circuit symbol for an LED**

Typical mistake

When drawing circuit diagrams, don't forget to draw a resistor in series with an LED.

Practise using the formula to calculate the resistor value for an LED. A common mistake is to forget to work out $(V_S - V_{LED})$ and to just insert V_{LED} or V_S into the equation.

Typical mistake

Don't confuse buzzers with loudspeakers. A buzzer will only buzz. A loudspeaker will not produce any sound unless it receives a sound waveform, e.g. music.

Driver: a subsystem used to boost a signal so that it can operate an output device.

Back emf: a high-voltage spike produced when motors, solenoids or relays are used.

- A solenoid can provide a pulling or a pushing force when it receives power. It produces a short reciprocating motion.
- Solenoids are used in electric door locks and in valves for liquids or gases.

Relay

- A relay is a switch (called the 'contacts') which is controlled by an electromagnet (called the 'coil').
- Relays allow a high-voltage (or a high-current) output transducer to be controlled from a low-voltage (or low-current) circuit.
- Relay coils are often designed to work at 6 V or 12 V.
- Relay contacts are often SPDT or DPDT switches and these will have a maximum voltage and current rating, e.g. 230 V, 10 A.

> **Exam tip**
>
> Don't forget to use a MOSFET and back emf diode when drawing circuits showing motors, solenoids or relays as output devices. Practise drawing these circuits as the symbols are quite complex.

Functions of mechanical devices/systems

`REVISED`

Rotary motion systems

- **Rotational velocity** is measured in revolutions per minute or revolutions per second.
- **Torque** is a turning force.
- In a simple mechanism, there is always a trade-off between rotational velocity and torque.
- A rotary mechanism can increase the torque but reduce the rotational velocity, or reduce the torque but increase the rotational velocity.

> **Rotational velocity:** the number of revolutions per minute (rpm) or per second (rps).
>
> **Torque:** a turning force.

Simple gear train

A simple gear train consists of two interlocking spur gears. See Section 2, Topic 5 for basic information on simple gear trains.

- The two gears rotate in opposite directions.
- The smaller gear rotates faster than the larger gear.

Velocity ratio

The **velocity ratio** is the factor by which a mechanical system reduces the rotational velocity.

> **Velocity ratio:** the factor by which a mechanical system reduces the rotational velocity.
>
> **Compound gear train:** more than one stage of gear train working together to achieve a high velocity ratio.

$$\text{Velocity ratio} = \frac{\text{rotational velocity of input}}{\text{rotational velocity of output}}$$

$$= \frac{\text{number of teeth on output (driven) gear}}{\text{number of teeth on input (driver) gear}}$$

This equation can be rewritten as:

(RV of input) × (number of teeth on input) = (RV of output) × (number of teeth on output)

Compound gear train

- Two or more sets of simple gear trains can work together to form a **compound gear train**.
- The overall velocity ratio is found by multiplying together the velocity ratios for each stage.

For a two-stage compound gear train:

Overall velocity ratio = (velocity ratio of stage 1) × (velocity ratio of stage 2)

- Each stage of a compound gear train reverses the direction of motion; therefore, for a two-stage compound gear train, the input gear and output gear rotate in the same direction.
- Compound gear trains are used when it is necessary for a mechanical system to have a large velocity ratio.

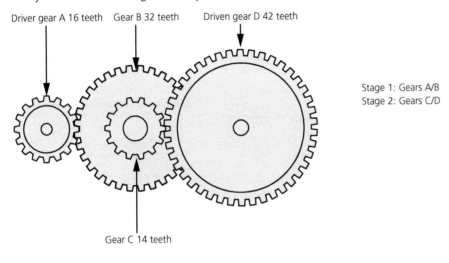

Driver gear A 16 teeth Gear B 32 teeth Driven gear D 42 teeth

Stage 1: Gears A/B
Stage 2: Gears C/D

Gear C 14 teeth

Gears B and C are locked together

Figure 2.19 **A compound gear train**

Pulley and belt drive

A pulley and belt drive system is similar to a simple gear train, but with the following differences:

- Pulley and belt systems transfer rotary motion between two shafts which can be separated by some distance.
- The input and output pulleys rotate in the same direction.
- They are quieter during operation than a gear system.

The smaller pulley rotates faster than the larger pulley, in the same way as a simple gear train.

For the purposes of calculations, it is the diameters of the pulleys which determine the velocity ratio.

$$\text{Velocity ratio} = \frac{\text{diameter of output (driven) pulley}}{\text{diameter of input (driver) pulley}}$$

The key equation is written:

(RV of input) × (diameter of input pulley) = (RV of output) × (diameter of output pulley)

Compound pulley system

- Compound pulley systems behave like compound gear trains, in that the overall velocity ratio is found by multiplying together the velocity ratio for each stage.
- In a compound pulley system, the input pulley and output pulley always rotate in the same direction, no matter how many stages there are.

Worm drive

- A **worm drive** consists of a worm screw (the input gear) and a worm wheel (the output gear).
- Worm drives achieve a very high velocity ratio.
- The velocity ratio is simply equal to the number of teeth on the worm wheel (for a single-start worm screw).
- The direction of motion is transferred through 90°.
- They are self-locking, which means that the worm screw can drive the worm wheel, but not the other way around.
- Worm drives are used in winches and lifts where the high velocity ratio and the self-locking feature are particularly useful.

Worm drive: a compact gear system which achieves a very high velocity ratio.

Bevel gears: a system to transfer the direction of rotation through 90°.

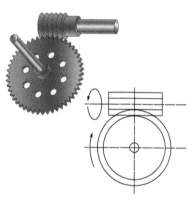

Figure 2.20 **A worm drive**

Bevel gears

- **Bevel gears** transfer rotary motion through 90°.
- The velocity ratio is calculated in exactly the same way as for a simple gear train.

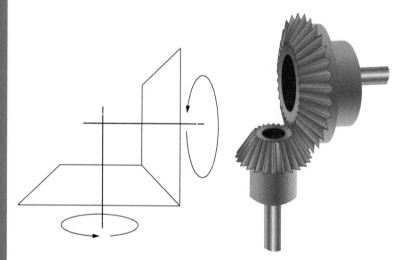

Figure 2.21 **Bevel gears**

Levers

A lever is a simple mechanism used to change forces and motion. See Section 2, Topic 5 for basic information on levers.

- A lever can amplify force, or it can amplify distance moved, but it cannot do both at the same time. If a lever amplifies the force, then it will reduce the distance moved, and vice versa.
- The **effort** is the input force and the **load** is the output force.
- The relative positions of the effort, load and **fulcrum** determine the class of lever.

Now test yourself answers at **www.hoddereducation.co.uk/myrevisionnotes**

In a first-class lever:

- the fulcrum is between the effort and the load
- effort and load move in opposite directions
- the exact position of the fulcrum determines whether the lever amplifies the force or amplifies the distance moved
- an example is a pair of scissors.

In a second-class lever:

- the load is between the effort and the fulcrum
- effort and load move in the same direction
- the force is amplified (load is greater than effort), but the distance moved is reduced
- an example is a wheel barrow.

In a third-class lever:

- the effort is between the load and the fulcrum
- effort and load move in the same direction
- the force is reduced (load is less than effort), but the distance moved is increased
- an example is a pair of tweezers.

Mechanical advantage

Mechanical advantage (MA) is the factor by which a system **increases** the force.

> Mechanical advantage: the factor by which a mechanical system increases the force.

$$\text{Mechanical advantage} = \frac{\text{output force}}{\text{input force}}$$

For a simple lever:

$$\text{Mechanical advantage} = \frac{\text{input arm length}}{\text{output arm length}}$$

- If a system amplifies a force, its MA is greater than 1.
- If a system reduces a force, its MA is less than 1.

Principle of moments

In a mechanical system:

Moment = force × perpendicular distance to fulcrum

The principle of moments states that, for a simple lever:

Effort × (input arm length) = load × (output arm length)

Ratchet and pawl

- This allows rotation in one direction only.
- It is used in turnstile entry gates, winches and lifts, and ratchet spanners.

Other systems

You will also need to understand rack and pinion, crank and slider and cams. For more details see Section 2, Topic 5.

7 Selection of materials and components

A designer is faced with several factors to consider when selecting materials and components. These include:

- function – what does the component do in the system? This might also include parameters such as size, resistance, material.
- aesthetic – is the look of the material important?
- environmental – is the product exposed to rain, dirt, sunlight or extreme temperatures?
- availability – does the supplier have sufficient stock for the batch to be manufactured?
- cost
- social, cultural and ethical issues.

Components and their functional benefits or limitations

REVISED

- The **rating** of a component is the maximum value of a specified quantity that it can handle.
- Operating a component beyond its rating is likely to damage it or reduce its life expectancy.
- Some components, such as batteries, have a limited lifespan.
- Designers should consider how products can be designed for maintenance, so that servicing can be carried out to extend the product's life.

Miniaturisation

- Prototype products made by hand are likely to be manufactured by entirely different methods in larger scales of production.
- Industrially-made **printed circuit boards (PCBs)** are usually double-sided to enable complex circuits to be achieved. Some PCBs are multi-layered.
- Industrial methods use **surface mount technology (SMT)** to attach the components to the PCB. The components do not have wire leads.
- SMT relies on high-speed pick-and-place robotic machines.
- The components are held in place by a solder paste, which turns into a permanent solder joint when the PCB is passed through a reflow oven.
- SMT allows complex, miniature circuits to be produced, which is essential for portable or wearable products.
- The PCBs are assembled at high speed, which reduces costs and increases the reliability of the end product.

Figure 2.22 **Filters in an air-conditioning unit need to be replaced regularly**

> **Rating:** the maximum specified quantity a component is designed to handle.
>
> **Printed circuit board (PCB):** a board with a pattern of copper tracks which complete the required circuit when components are soldered on.
>
> **Surface mount technology (SMT):** the industrial method of mounting miniature components onto a PCB using robotic machines.

Cultural, social, ethical and environmental responsibilities of designers

REVISED

- The Restriction of Hazardous Substances (RoHS) directive reduces the use of materials such as lead, cadmium and mercury in electronic equipment.
- Hazardous substances can be found in batteries, paints, solder and polymers.

Now test yourself answers at www.hoddereducation.co.uk/myrevisionnotes

- Designers have an ethical responsibility for the working conditions of the people who manufacture the products.
- Designers have a responsibility to ensure that the materials used in the products are ethically sourced.
- The Ethical Trading Initiative (ETI) promotes workers' rights around the globe.

8 Stock forms, types and sizes

Metals

REVISED ☐

- It is important to consider the common stock forms and sizes of metal when designing and making metal components. This will save you time and effort as metal is a hard material to cut and shape.
- Some of the most popular stock forms of metal are shown in Figure 2.23.

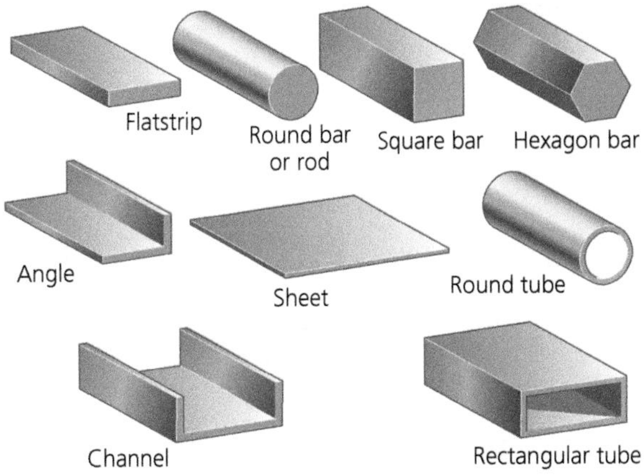

Standard stock electronic component sizes

- Most electronic components for prototyping have leads which will fit through circuit board holes on a 0.1 inch (2.54 mm) grid.
- Many components, including resistors and capacitors, are manufactured in certain values, called preferred values.

Figure 2.23 **Standard metal forms**

Dual-in-line standard for electronic ICs

- **Dual-in-line (DIL)** ICs have pins which fit the standard 0.1 inch (2.54 mm) spacing.
- There is a notch in one end of the IC and pin 1 is to the left of this. Pin 1 is sometimes identified by a small dot.
- The pins are numbered anticlockwise around the case.

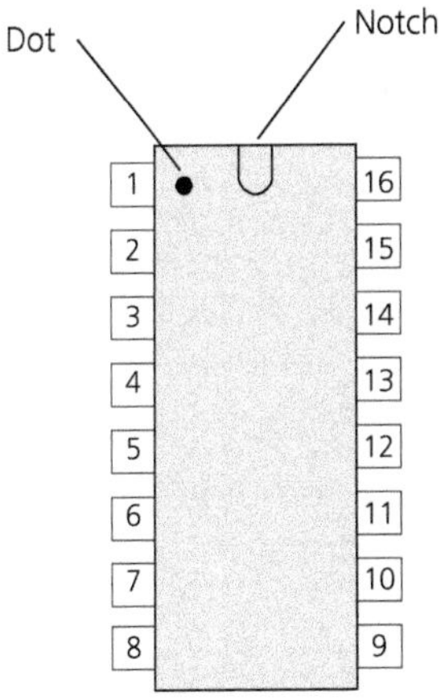

Manufactured boards and polymers

For details of stock forms for manufactured boards, see Section 4, Topic 8. For details of stock forms for polymers, see Section 4, Topic 8.

Figure 2.24 **DIL pin numbering**

Calculate the costs involved in the design of engineering products

REVISED ☐

- The costs of the components used in a project can be found in the supplier's catalogue.
- Every component used in the project, including wires, resistors, nuts and bolts, needs to be costed.
- Many suppliers offer price breaks if several items are bought – see Table 2.2, for example.

Table 2.2 **Examples of price breaks for bulk buying of components**

10+	100+	1000+
£0.121	£0.080	£0.051

- The '10+' means that the minimum quantity the supplier will sell is ten, at a cost of 12.1p each.
- You would have to spend a minimum of £1.21 to buy ten LEDs.
- If you buy 100 or more LEDs, then the price drops to 8.0p each,
- The price drops to 5.1p each if you buy 1000 or more LEDs.

Costs of materials and finishes

To calculate the cost of materials that you use in a project:

1 Find the cost of a full length (or a full sheet).
2 Calculate the cost per unit length (or unit area).
3 Measure the length (or calculate the area) you have used.
4 Calculate the cost.

Remember to include the costs involved in preparing the surface, such as sanding wood or cleaning the surface of metal before applying paint.

9 Manufacturing to different scales of production

Production

REVISED

One-off (bespoke) production

- **One-off production** is expensive because the designer will not receive rewards from any future sales.
- Products are often hand-made, requiring a high level of skill.
- **Rapid prototyping** may be used (see below).

> One-off production: process used when making a prototype product.

Batch production

- This method of manufacturing has a predetermined number of items.
- There is no limit on batch size, but production will continue until the entire batch is completed.
- Manufacturers organise a production system to suit their machinery and their workforce.

Mass production

- This is used for the manufacture of commonly used items, such as screws and batteries.
- Specialist factories produce items quickly at high speed.
- Repeatability and consistency of the items are achieved.
- In **continuous flow production**, items are produced 24 hours a day.
- The cost per item is usually extremely low.

> Continuous flow production: identical products are being constantly made due to the high demand.

Just-in-time (JIT) manufacturing

- This achieves manufacturing efficiency by ensuring that materials are ordered to arrive just before they are needed for a batch manufacture.
- The finished products are shipped out soon after completion, so there are no storage costs.
- JIT manufacturing achieves efficient flow through a factory.

10 Specialist techniques

The use of CAD/CAM in production

- 2D and 3D CAD software can link directly to CAM machines such as a laser cutter, CNC router, plasma cutter, vinyl cutter or CNC lathe.
- CAM machines achieve accurate and repeatable cuts in materials.
- A 3D CAD design can be quickly realised on a 3D printer. This is called rapid prototyping.
- PCB design software is used to route the PCB tracks in complex circuit designs and to test the function of the design.

Jigs and devices to control repeat activities

- A **jig** guides a tool so that it cuts in the right place.
- A fixture holds the material in exactly the right place so it can be processed accurately.
- A template is used for repeated marking out onto materials.
- A former is used to achieve a precise bend angle.

> **Jig:** mechanical aid used to manufacture products more efficiently.

Wastage/addition

You should be familiar with the application and use of the following tools:

- Marking-out tools include steel rule, try-square and marker/scriber.
- Bench hook, vice and G clamp for holding work during fabrication.
- Hand-cutting tools including tenon saw, coping saw and hacksaw, and tin snips and nibbler for sheet metal.
- Machine-cutting tools such as scroll saw and bandsaw.
- Shaping tools such as file, needle file and sanding machine.
- Drilling tools including cordless drill, bench drill, twist drill, Forstner bit, flat bit, hole cutter and cone cutter.

See Section 4, Topic 10 for more details on these tools and their uses.

Deforming/reforming

You should be familiar with the following methods of **deforming/ reforming**:

- Bending polymers – line bending, drape forming.
- Vacuum forming – shape and design of former, draft angle, vent holes.

Hot/cold working of metals

You should be familiar with the following methods:

- Cold working – folding, punching, rolling. Cold working can cause metal to become hard and brittle.
- Hot working – **annealing** to increase **ductility** to prevent cracking when bending.
- Casting – industrial casting commonly uses iron, brass or aluminium. Casting in school usually users pewter, an alloy which melts around 230°C.

> **Deforming:** changing the shape of a material by applying force, heat or moisture.
>
> **Annealing:** a heat treatment to reduce cracking when bending metal.
>
> **Ductility:** the property of a material to be able to be permanently stretched out without cracking.

Other processes

Other processes include the following.

Turning:

- Using a centre lathe for metal.
- Using a wood lathe.

Laser cutting:

- Setting laser power and speed of cut.
- Focusing the laser.
- Need for material to be flat.

3D printing:

- Materials used include polymers, metals, ceramics or food.
- Used for the fast production of prototype parts.
- Useful for bespoke production.

> **Turning:** a method of producing cylinders and cones using a centre lathe.

Assembly and components

REVISED

Temporary joining methods

Nuts and bolts:

- This requires a clearance hole to be drilled through both parts.
- There is bolt identification labelling, e.g. M4 20.
- A washer is often used under the nut to spread the pressure over a greater surface area.

Machine screws:

- These are bolts that are threaded along their entire length.
- They screw into a pre-threaded hole in a metal part.
- A tap is used to cut a thread into a pre-drilled pilot hole in metal.

Self-tapping screws (for metal) and wood screws:

- These require a clearance hole through one part and a pilot hole into the other part.

Permanent joining methods

Pop rivets for sheet materials:

- This requires a clearance hole through both parts.

Adhesives:

- It is important to choose the correct adhesive depending on the materials to be joined.
- Surfaces need to be cleaned and prepared.
- Types of adhesive and their application include polyvinyl acetate (PVA), contact adhesive, Tensol, epoxy resin, hot melt.

Soldering:

- This is used for making permanent electronic joints between components and on circuit boards.
- Solder is an alloy which melts around 220 °C.
- Flux is used to clean the joint.

Brazing:

- This is carried out at a higher temperature than soldering.
- It can be used to join steel, aluminium, copper and brass, for example.
- A filler metal is used in the join.

Welding:

- This is the strongest method of joining metals.
- The metals melt and fuse together.

Temporary and permanent fixings of circuits

- Electronic components, printed circuit boards and batteries all need to be held securely within a product.
- Heavy components such as batteries can cause a lot of damage to other parts if they move around inside the product casing, and short circuits can occur if parts touch each other.
- A well-designed product will address these issues in the early stages of the design.

11 Surface treatments and finishes

Surface finishes applied to electronic devices

Casings are used to protect electronic systems and to improve their aesthetic appearance.

Polymers:

- Casings made from polymers are likely to be **self-finished**, so they do not require any additional surface finishes.

Metals:

- Steel will rust unless it is protected with a finish. Finishes for steel include a film of oil or paint.
- Before applying paint, the surface must be cleaned and a coat of **primer** applied before the final coat(s) of paint in the desired colour. Paint can be applied by brush or spray.
- Aluminium and brass do not rust, but they do oxidise, so they are often polished and a clear lacquer is applied to maintain an attractive look.
- Brushed and lacquered aluminium is a popular finish.

Woods:

- A **preservative** can be used to prevent wood decaying when used outdoors.
- Wood can be painted. The surface is sanded, then a primer coat is applied. Further coats are applied, lightly sanding between each coat.

Other considerations:

- User instructions or product safety information may be required. Images and diagrams could be used to make these clear.
- Controls and indicators such as the on/off switch should be labelled. Icons could be used instead of words and scale markings may be useful, for example for a volume control.

> **Exam tip**
>
> Questions relating to manufacturing will score the highest marks if you use the precise names of tools, processes and materials rather than generic words such as 'saw', 'moulding' or 'plastic'.

> **Self-finished**: a material that does not require the application of a finish to protect it or improve its appearance.
>
> **Primer**: the base coat of paint applied straight to the material surface.
>
> **Preservative**: a chemical treatment applied to wood to prevent biological decay.

- PCBs used in extreme conditions will need to be protected from extreme temperatures, moisture or vibration. This can involve the PCB being coated in resin, which protects the PCB but prevents parts being replaced if a fault develops.
- Any company logos, the product name and trademarks may also need to be added.

Figure 2.25 **Labelling on a control panel**

Powder and polymer coating of metals

Powder coating is a form of painting:

- The metal is cleaned by **shot-blasting**.
- A dry, coloured polyurethane powder is sprayed onto the metal, using an electrostatic charge to encourage the powder to stick.
- The metal is baked in an oven, causing the polymer powder to melt and fuse into a smooth coating.
- Powder coating is used on fridges, washing machines, bicycle frames, etc.

Dip-coating is a similar process:

- The metal part is heated.
- It is then plunged into a **fluidised bath** of coloured polymer powder.
- The powder melts and fuses to the metal.
- Dip-coating is used on tool handles, cupboard door handles, coat hooks, etc.

> **Shot-blasting:** using grit, fired at high pressure, to clean a surface by abrasion.
>
> **Fluidised bath:** blowing air through a powder to cause it to behave like a fluid.

Now test yourself

1 (a) Draw a system diagram, based around a microcontroller, for a night light which turns on an LED when the light level drops below a threshold, or when a switch is pressed. [4]
   (b) Add labels to the system diagram from part (a) to identify the analogue and digital signals in this system. [3]
   (c) Calculate the LED resistor value if the high output from the microcontroller is 5 V, the LED has a voltage drop of 2.2 V and the LED current is 8 mA. [3]
   (d) Explain why a resistor needs to be used with an LED. [2]
2 Draw the circuit symbols and explain the action of the following electronic components:
   (a) Op-amp [2]
   (b) MOSFET [2]
   (c) Relay. [2]
3 Five spur gears are available with the following number of teeth:
   15t 20t 30t 40t 60t
   Draw a diagram to show how four of these gears can be used to make a compound gear train with a velocity ratio of 6. [4]
4 Draw three labelled diagrams to show the differences between the three classes of lever. [3]

5 Describe one way in which legislation has impacted on how society disposes of unwanted or obsolete products. [3]
6 Describe the benefits offered by the use of surface mount technology (SMT) in electronic products. [3]
7 Describe, with examples, two reasons why electronic systems and components are often placed in a casing. [4]
8 Metal bars are available in a variety of stock profiles. Sketch and name three different metal bar stock profiles. [3]
9 Give two reasons why one-off production is expensive. [2]
10 Use sketches and notes to describe the steps involved in casting a pewter circle, of about 40 mm diameter. [8]
11 Describe the differences between brazing and welding as methods of permanently joining metal plates. [4]
12 Describe the powder coating process for painting steel wheels. [4]

Exam practice

1 A coffee machine uses a thermistor to sense the temperature of the hot water and a microcontroller to control the function of the machine.
   (a) Complete this sentence: As the temperature of an NTC thermistor rises, its resistance
   .... [1]
   (b) Sensors can be classified into analogue or digital types. Explain why a thermistor is an analogue sensor. [2]
   (c) Describe two advantages to a designer of using a microcontroller in a product. [4]
   (d) The temperature control system for the coffee machine uses feedback.
      (i) Explain what is meant by feedback in a control system. [2]
      (ii) Give one advantage of using feedback in a control system. [1]
2 A toy car is propelled by an electric motor. The motor produces rotary motion at high speed.
   (a) Apart from rotary motion, identify two other types of motion. [2]
   (b) Draw a labelled diagram to show a mechanical system which would reduce the rotary speed produced by the motor. [3]
   (c) Describe two functional reasons for using levers in mechanical systems. [2]
3 An outdoor security light switches on for 45 seconds when movement is detected.
   (a) Give one reason why LEDs are the preferred source of light in modern security lights. [1]
   (b) The LED used in this system has a voltage drop of 11.4 V, and a 33 ohm resistor is used with the LED.
      The power supply voltage is 24 V.
      Calculate the current flowing (in mA) through the LED. [3]
   (c) The outdoor security light is housed inside a box made from mild steel sheet.
      (i) Describe two problems a designer must consider when developing a security light for outdoor use. [2]
      (ii) Use sketches and notes to describe a manufacturing and finishing method for producing a prototype box for an outdoor security light from mild steel sheet. Identify tools and processes and materials used. [6]
   (d) Discuss ways in which an outdoor security light can be developed as part of a sustainable design strategy. [4]

ONLINE

3 Fashion and textiles

1 Natural, synthetic, blended and mixed fibres

- **Fibres** are very fine hair-like structures that are spun (or twisted) together to make yarns.
- Yarns are then woven or knitted together to create textile fabrics.
- Fibres have different properties and characteristics that affect what they can be used for.

> **Fibre:** a fine hair-like structure.

Natural polymers

REVISED

- **Natural polymers** come from natural sources: plants (**cellulosic**) and animals (**protein**).
- They are sustainable and **biodegradable**.
- The sources of natural fibres are:
  - plant polymers: extracted from the stem or seeds of plants, and cellulose-based fibres extracted from, for example, wood pulp
  - insect polymers: extracted from insects
  - animal polymers: fibre from the hair or fleece of animals such as sheep, goats (mohair, cashmere), rabbit (angora), alpaca camel, etc.

> **Natural polymers:** polymers that are sourced from plants and animals.
>
> **Cellulosic fibres:** natural fibres from plant-based sources.
>
> **Protein fibres:** natural fibres from animal-based sources.
>
> **Biodegradable:** a material that will decompose into the Earth.

Table 3.1 Properties and common uses of natural polymers

Fibre	Source	Properties	Uses
Cotton	Plant	Absorbent, strong, cool to wear, hard wearing, creases easily, smooth, easy to care for, flammable, can shrink	Clothing, sewing and knitting threads, soft furnishings
Linen	Plant	Strong, cool to wear, absorbent, hard wearing, creases very easily, has a natural appearance, handles well, flammable	Lightweight summer clothing, soft furnishings, table linen
Hemp	Plant	Absorbent, non-static, anti-bacterial, naturally lustrous, strong	Clothing, carpets and rugs, ropes, mattress filling
Jute	Plant	Very absorbent, high tensile strength, anti-static	Bags, sacking, carpets, geotextiles, yarn and twine, upholstery, clothing but to a lesser extent
Bamboo	Plant	Soft, fine and lustrous, non-irritant, absorbent, anti-static, crease resistant, biodegradable, high tensile strength, UV resistant, antimicrobial	Dresses, shirts, trousers, socks, activewear, sheets and pillow cases
Soya	Plant	Soft and smooth, lightweight, lustrous, absorbent, crease and shrink resistant, UV resistant, antibacterial, biodegradable	Clothing including dresses, cardigans and jumpers, soft furnishings

Fibre	Source	Properties	Uses
Silk	Insect	Absorbent, comfortable to wear, can be cool or warm to wear, strong when dry, has a natural sheen, creases	Luxury clothing and lingerie, knitwear, soft furnishings
Wool	Animal	Warm, absorbent, low flammability, good elasticity, crease resistant	Warm outerwear including coats, jackets and suits, knitwear, soft furnishings including carpets and blankets

Manufactured polymers

- Manufactured or **synthetic polymers** are artificial fibres derived from oil, coal, minerals or petrochemicals
- The fibres (known as monomers) are joined together by **polymerisation**, then spun into yarns before being woven or knitted into fabrics.
- An advantage of **synthetic** fibres and yarns is that they can be engineered for specific purposes.
- Most synthetic polymers are non-biodegradable and from unsustainable sources.

> **Synthetic polymers:** polymers that are sourced from crude oil.
>
> **Polymerisation:** chemical reaction that causes many small molecules to join together and form a larger molecule; the blending of different monomers to create a specific polymer.
>
> **Synthetic:** derived from petrochemicals or manmade.

Table 3.2 **Properties and common uses of synthetic polymers**

Fibre	Properties	Used for
Polyester	Strong when wet and dry, flame resistant, thermoplastic, hard wearing, poor absorbency	A very versatile fabric used throughout textiles
Nylon (polyamide)	Strong and hard wearing, melts as it burns, thermoplastic, good elasticity, poor absorbency	Clothing, carpets and rugs, seat belts and ropes, tents
Polypropylene	Thermoplastic with a low melting point, strong, crease resistant, non-absorbent, resistant to chemicals, hard wearing and durable	Engineered for specific uses to include carpet backing, sacks, webbing, twine, fishing nets, ropes, some medical and hygiene products, awnings, geotextiles
Acrylic	Strong except when wet, thermoplastic, burns slowly then melts, poor absorbency	Knitwear and some knitted fabrics, fake fur products including toys, upholstery
Elastane, Lycra	Very elastic and stretchy, lightweight, strong and hardwearing	Clothing but particularly swimwear and sportswear where stretch, comfort and fit are critical
Aramid fibres	Engineered for strength and heat resistance, no melting point, five times stronger than nylon, resistant to abrasion, low shrinkage, ease of care	Flame-resistant clothing, protective clothing, accessories, armour, geotextiles, aeronautical industry, ropes and cables, high-risk sports equipment

Regenerated polymers

- Regenerated fibres are made from plant cellulose extracted from wood pulp from eucalyptus, pine or beech wood and cotton linters.
- A chemical solution is added during the extraction process so they are part-natural, part-artificial.

Table 3.3 **Properties of regenerated polymers**

Fibre	Source	Properties	Uses
Viscose (Rayon)	Plant/ chemical	Blends well with other fibres, breathable, drapes well, excellent colour retention, highly absorbent, relatively light, comfortable, soft to the skin, moderate strength and abrasion resistance	Linings, shirts, blouses, shorts, dresses, sportswear, coats, jackets and other outerwear
Acetate	Plant/ chemical	Drapes well, creases easily, prone to static, biodegradable	Clothing and furnishings, a cheaper alternative to silk
Lyocell (Tencel™)	Plant	Biodegradable, strong, soft, absorbent, resists creasing	Shirts, suits, skirts, leggings, household linen

Microfibres

REVISED

- **Microfibres** can be natural or manufactured.
- They are up to 100 times finer than a human hair.
- Microfibres can be engineered to create fabrics with specific qualities and functions, such as lightweight, strong, crease resistant, soft.
- Products made from microfibres include sportswear, underwear and high-performance garments.

> **Microfibre:** an extremely fine specially engineered fibre.

Table 3.4 **Examples of microfibres**

Microfibre	Source	Properties	Uses
Tactel® Polyamide (nylon) fibre	Manufactured	Hardwearing, quick drying, crease resistant	Often blended with cotton or linen Used mainly for underwear and active wear
Modal	Manufactured	High strength, good absorbency	Often blended with cotton or polyester Suitable for use in underwear

Blending and mixing fibres

REVISED

Fibres are often mixed or blended together to improve the properties of the **yarn** or fabric to improve:

- the quality, for example to make it stronger or easier to care for
- the appearance, such as the texture, tone or colour
- functionality, for example to improve the handle of the fabric so that it drapes better
- the cost of the yarn or fabric, for example by blending an inexpensive yarn with an expensive yarn to reduce the overall cost.

> **Yarn:** spun thread used for knitting, weaving or sewing.

Mixed fibres

- Fibres are mixed together by adding yarns of different fibres during the production of the fabric.
- One yarn is used for the warp yarns that run along the length of the fabric and a different one for the weft yarns that are combined with the warp yarns across the fabric.

Blended fibres

- Blended fibres are two or more different fibres spun together to make a single yarn.

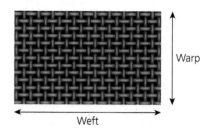

Figure 3.1 **In a mixed-fibre fabric, the warp yarns would be one fibre and the weft yarns would be a different fibre**

Now test yourself answers at www.hoddereducation.co.uk/myrevisionnotes

- The most common blend is polyester cotton.
- Cotton is absorbent, soft and strong.
- Polyester is hard wearing, quick drying and elastic.
- The combined properties create a versatile fabric that is comfortable and cool to wear like cotton, but with the added features of being quick drying and crease resistant.

Exam tip

By making things in school or at home using the different types of materials you will learn their different characteristics and properties.

2 Woven, non-woven and technical textiles

Woven fabrics

REVISED

- Weaving is done on a loom using **warp** and **weft** yarns.
- Warp yarns run along the length of fabric, called the **straight grain**.
- Weft yarns run horizontally across the fabric, called the **cross grain**.
- Weft yarns interlock with warp yarns in different formations, creating variations in types of **weave**.
- The **selvedge** is the factory-finished edge of the fabric.
- Different weaves create fabrics with different textures, patterns and strength.

Warp: yarns that run along the length of the fabric.

Weft: yarns that run across the fabric.

Straight grain: indicates the strength of the fabric in line with warp yarns.

Cross grain: runs horizontally across the fabric in line with the weft yarn.

Weave: the pattern woven in the production of fabric.

Selvedge: the sealed edge of the fabric.

Table 3.5 **Characteristics of woven fabrics**

Woven fabric	Characteristics
Plain weave	The simplest structure and most commonly used weave Stable, strong and gives an even surface on both sides of the fabric Variations can be achieved through different thickness and textures of yarn, different colour combinations in the yarns and how closely the yarns are packed together
Twill weave	Recognised by the diagonal lines created by the weave which adds strength to the cotton fabric Variations include herringbone and chevron Twill weave produces a strong, heavy and more durable fabric Used for denim jeans
Satin weave	A smooth, shiny, lustrous appearance created by the floating yarns in the weave A disadvantage is that it snags easily due to the structure of the floating yarns
Pile weave	Has a raised surface formed by tufts or loops in the weave which can be cut as in velvet or left as in towelling Pile weave fabrics are hard wearing because of the thickness created by an extra loop in the yarn Pile fabrics such as corduroy have a directional surface so all pattern pieces have to be laid and cut in the same direction to avoid shading

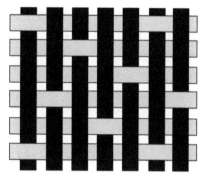

Figure 3.2 A plain weave structure   **Figure 3.3** A twill weave structure   **Figure 3.4** A satin weave structure

Knitted fabrics

- Knitted fabrics are made by creating a series of loops in the yarns that interlock together.
- Knitted fabric is easy to stretch and warmer to wear as the loops trap body heat.
- There are two types of knitted fabric: warp knitting and weft knitting.

Table 3.6 **Knitted fabrics**

Weft knit fabrics	Warp knit fabrics
Made from a single continuous yarn, constructed from horizontal rows of interlocking loops	Made on automated machines from multiple yarns that interlock vertically
Can be hand-made or constructed on industrial machines	Identical on both sides
Have an obvious right and wrong side	Do not run or unravel
Can snag	More flexible than weft knits
If part of the yarn is damaged, cut or pulled, they can unravel	Have some stretch
Can stretch easily	Hold their shape well and can be cut to shape when making products
Can lose shape	

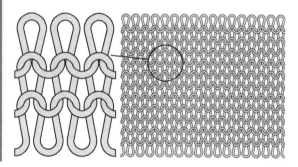

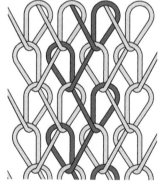

Figure 3.5 The weft knitting structure has a series of loops made from a single yarn that interlocks horizontally

Figure 3.6 Warp knitting consists of multiple yarns interlocking vertically

Bonded fabrics

- Bonded fabrics are made directly from fibres, which makes them cheaper to use.
- Pressure and heat or adhesives are applied to a web of fibres to bond them together.
- Bonded fabrics have limited uses but are often used in disposable products such as surgical masks.

Now test yourself answers at www.hoddereducation.co.uk/myrevisionnotes

Laminated and coated fabrics

- Fabrics are laminated to combine the performance characteristics of the fabrics used to make a superior fabric. Examples include neoprene as used in wetsuits.
- Laminated fabrics are either:
  - held together with an adhesive, or
  - either a polymer film or a layer of foam is heated and pressed onto the fabric it is to be joined to.

Felted fabrics

- Felted fabrics are non-woven fabrics made by applying pressure, moisture, heat and friction to staple fibres to bind them together.
- They are usually made from wool or acrylic fibres. They are easy to cut and will not fray but can stretch out of shape when wet. They are ideal for craft projects.

Technical textiles

Technical textiles are engineered to have specific performance characteristics for a particular purpose or function. Examples include:

- Gore-Tex® – consists of three or more fabrics laminated together with a breathable **hydrophilic membrane** in the middle. Warm air and tiny droplets of moisture from perspiration permeate out through the breathable membrane but moisture from larger rain droplets and wind cannot enter. When used in high-performance clothing and footwear, it helps to regulate body temperature by maintaining a constant temperature through allowing the flow of air in and out.
- Permatex – a weatherproof, breathable membrane used for linings and outer fabrics for high performance clothing, footwear, industrial wear and sportswear.
- Sympatex – another example of a hydrophilic membrane which is extremely thin and dense so that the wearer does not feel even the slightest effect from high wind.

> **Typical mistake**
>
> Technical fabrics such as Gore-Tex, which include a hydrophilic membrane, appear 'breathable' by allowing moisture to pass through. These technical fabrics should not be confused with other 'absorbent' fabrics such as cotton that simply soak up moisture and are technically not breathable.

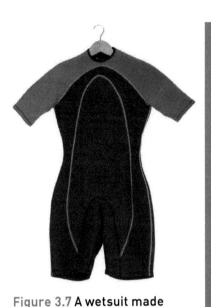

Figure 3.7 A wetsuit made from laminated neoprene

Figure 3.8 Hand-made felt is made by applying pressure, heat, moisture and friction

> **Hydrophilic membrane:** a solid structure that stops water passing through but at the same time can absorb and diffuse fine water vapour molecules.

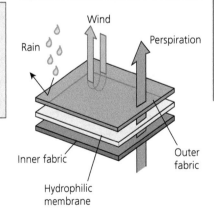

Figure 3.9 Gore-Tex is an example of a laminated fabric and is widely used for high-performance clothing.

Geotextiles

- **Geotextiles** are woven or bonded, synthetic or natural, permeable fabrics made originally for use with soil, with the ability to filter, separate, protect and drain.
- Geotextiles have many uses and applications in civil engineering, road and building construction and maintenance, for example control of coastal erosion and drainage, roofs such as on the Eden Project in Cornwall and control of embankments on the sides of roads.

Nomex

- Nomex® is an aramid synthetic fibre that is primarily used where resistance to heat and flames is essential, for example in firefighters' protective clothing, oven gloves and fire-resistant insulation on buildings.
- It is an extremely strong fabric and can withstand exposure to the most extreme conditions.

Carbon-fibre reinforced polymer (CFRP)

- CFRP consists of woven carbon fibre strands encased in a polymer resin.
- Carbon fibre strands have a very high tensile strength and the polymer resin is lightweight and rigid, creating a high-performance engineered material. It is often used in sports equipment such as a tennis racquet where strength-to-weight ratio is important.

Kevlar

- Kevlar® is a lightweight, flexible and extremely durable **aramid fibre** that has excellent resistance to heat, corrosion and damage from chemicals and has a high tensile strength-to-weight ratio.
- It is often used in protective clothing such as police body armour, where the fibre is woven in a lattice that provides protection against knife attack.

Biosteel

- Biosteel is created from spider silk, which is extracted from milk from genetically modified goats that have been transplanted with the spider-silk gene and spun into spider-silk thread.
- Spider silk is from a group of protein-based materials known as biopolymers, hence the name 'biosteel'.
- Biosteel has excellent extensibility and can stretch up to 20 times its normal length without breaking. It is considered comparable in tensile strength to steel and could replace Kevlar for some of its uses if it can be produced in a commercially viable way.

3 Thermoforming and thermosetting polymers

- The categorisation and properties of polymers used in fashion and textiles can be found in Section 3, Topic 1.
- Polymers can be split into two groups – thermoforming and thermosetting polymers.

> **Geotextiles:** textiles associated with soil, construction and drainage.

Figure 3.10 A geotextile is used on the roof of the Eden Project in Cornwall

> **Aramid fibre:** a non-flammable heat-resistant fibre at least 60 times stronger than nylon.

Now test yourself answers at www.hoddereducation.co.uk/myrevisionnotes

Thermoforming polymers

- **Thermoforming polymers** can be softened by heating and moulded into almost any shape using a wide variety of processes such as pleating and moulding.
- Once the desired shape has been achieved, the polymer cools and maintains its new shape.
- Fabrics such as polyester, nylon, polypropylene and acrylic are thermoforming polymers that can be pleated, moulded and easily shaped.
- Nylon and polyester lend themselves to the process of pleating and moulding as they are pliable above a certain temperature but will not melt.
- Other thermoforming polymers such as polythene, polystyrene and polyvinyl chloride (PVC) are used in some textile products.

> **Thermoforming polymer:** a polymer that can be reheated and reformed.
>
> **Thermosetting polymer:** a polymer that cannot be reformed with heat.

Thermosetting polymers

- Thermosetting polymers can be shaped and formed once only. They cannot be reheated or reformed once they have been formed and cooled, which means that they cannot be recycled.
- Components such as plastic clips, buckles and buttons are made from thermosetting polymer because it is highly malleable and easily moulded into different shapes.

Table 3.7 Examples of thermoforming and thermosetting polymers

	Type of polymer	Properties	Uses
Polypropylene (PP)	Thermoforming	Semi-rigid Translucent Good chemical resistance Tough/strong Hardwearing/durable Crease resistance Good heat resistance/low melting point	Carpet backing Sacks and bags Webbing Fishing nets Ropes and twine Some medical and hygiene products Awnings Geotextiles
Polythene (PE)	Thermoforming	Tough Flexible Easily moulded High tensile strength	Carrier bags and mail sacks Bin liners Dust sheets Garment covers
Polystyrene (PS)	Thermoforming	Lightweight Good impact strength	Styrofoam for block printing/craft projects Beads (filling in beanbags) Fillings in packaging
Polyvinyl chloride (PVC)	Thermoforming	Flexible or rigid Dense Good tensile strength Water resistance	Shopping bags Purses Toiletry/make-up bags Raincoats, hats, skirts and jackets Boots and shoes

4 Modern and smart materials

Interactive textiles

REVISED

- **Interactive textiles** or integrated textiles are those in which electronic devices and circuits are integrated or embedded into textile fabric and clothing to interact and communicate with the wearer.
- Conductive fibres and threads developed from carbon, steel and silver can be woven into textile fabrics and made into clothing or conductive threads can be sewn into a product to connect a circuit.
- Uses include heart rate monitors, performance monitors for athletes, GPS tracking systems, heating and flexible solar panels and communications devices such as mobile phones.

> **Interactive textiles:** fabrics that contain devices or circuits that respond and react with the user.

Photochromic pigment

- Photochromic pigments or dyes change colour in response to changes in light, for example sunglasses can change colour in response to UV radiation.

Thermochromic pigment

- Thermochromic pigments or dyes change colour in response to a change in heat and can be engineered to specific heat ranges.
- Thermochromic dyes can be used in medical dressings; a change in colour indicates heat which could mean an infection is present.

Figure 3.11 Flexible solar panels can now be integrated into textile fabrics

Micro-encapsulation

- Micro-encapsulation is a process of applying microscopic capsules to fibres, fabrics, paper and card.
- The capsules can contain vitamins, therapeutic oils, moisturisers, antiseptics and anti-bacterial chemicals which are released through friction.

Biomimetics

Ideas for new materials and products can be inspired by nature. For example, the hoop-and-loop fastening known as Velcro® was invented by Swiss engineer George de Mestral. While out hunting, he noticed that the tiny hooks of the cockleburs stuck to his clothes and to his dog's fur. This inspired him to investigate further, and Velcro was the result.

Microfibres

- Microfibres are extremely fine, lightweight synthetic fibres, usually polyester or nylon, that are significantly finer than a human hair.
- Useful properties include excellent strength-to-weight ratio, water-resistance and breathability.
- Fabrics made from microfibres are used throughout the textiles industry, from clothing to cleaning cloths.

Phase changing materials

- **Phase-changing materials (PCMs)** change from one state to another, with the ability to absorb, store and release heat over a small temperature range by changing from liquid state to solid and vice versa.

> **Phase-changing materials:** droplets encapsulated on fibres and materials that change between liquid and solid within a temperature range.

- PCMs absorb energy during the heating process (returning to liquid) and release energy to the environment during cooling (returning to solid).
- In cold-weather clothing, PCMs encapsulated into the fabric allow body heat to be stored within the fabric and then released when needed.

Sun-protective clothing

- Clothes made from tightly woven or knitted fabrics are the most effective at blocking out the sun's harmful UV rays because the gaps between the yarns are significantly reduced, which prevents the rays from getting through.
- Elastane fibres used in fabric reduce the spaces even further, making these even more efficient and protective.

Rhovyl

- Rhovyl® is a non-flammable, synthetic fibre which is crease resistant, has good thermal and acoustic properties, is anti-bacterial and comfortable to wear.
- The construction of the fibre gives the fabric the ability to wick away moisture such as perspiration through the fabric. It also dries quickly, meaning it does not retain odours, making it ideal for socks.

Breathable fabrics

For details of breathable fabrics such as Gore-Tex, see Section 3, Topic 2.

Now test yourself
TESTED ☐

1 What is a microfibre? [3]
2 Explain why an athlete might favour clothing made from Rhovyl fibre. [3]
3 Explain why thermosetting polymers cannot be recycled. [2]
4 Describe two examples where conductive threads have been used in textile products and the impact of this technology on the user. 2 × [2]
5 Thermochromic dyes are used in wound dressings. Explain the benefits to the patient. 2 × [3]
6 Explain why pleats can be created in a polyester fabric. [3]
7 Describe two reasons why the aramid fibre Nomex is used in protective clothing. 2 × [2]

Exam tip

Be able to fully explain how smart, composite and technical textiles are used in the design and manufacture of products. Know their specific properties and what makes each suitable for a particular purpose in a product.

5 Sources, origins, physical and working properties

Fabric
REVISED ☐

Choosing the most suitable fabric for a product is vital. Factors that need to be considered include:

- the source of the fibre and its properties
- how the fibre has been spun and made into yarn
- how the fabrics have been constructed from yarns or fibres
- applied finishes.

For information about weaving, knitting, bonding, laminating and felting, see Section 3, Topic 2.

Fabric specification

- A **fabric specification** sets out the fabric requirements for a product for a specific end purpose.
- The physical and working properties of fibres, **fabric construction** and applied finishes should be carefully considered before selecting an appropriate material to use.

> **Fabric specification:** sets out the requirements of the fabrics needed for a product.
>
> **Fabric construction:** the way a fabric has been made.

Fibres

REVISED

- Fibres are the raw materials of textiles and come from natural or manufactured sources.
- They are made from chemical units called polymers, formed from smaller single units called **monomers** which link together, creating long chains.
- Fibres are classed as:
  - long continuous **filaments**, for example polyester, nylon and acrylic
  - short staple fibres, for example cotton, linen and wool.
- The shape of fibres affects how the fibre feels or its **handle** (softness) and **lustre** (shine).
- Fibre blends bring together fibres of different types to improve functionality, cost or appearance of the blended/mixed fibres.
- Common fibre blends include polyester cotton, cotton and elastane, wool and acrylic, silk and viscose.

> **Monomer:** a molecule that can be bonded to others to form a polymer.
>
> **Filament:** a very fine and slender thread.
>
> **Handle:** how a fabric feels when handled.
>
> **Lustre:** a gentle shine or soft glow.
>
> **Crimp:** the waviness in a fibre.

Table 3.8 **Structure and description of fibre types**

Fibre	Description
Cotton	Short staple fibres with a slight twist
	Smooth surface prevents air being trapped – poor insulator
	Inner cavity allows moisture to be absorbed
Linen	Short staple fibres
	Smooth surface prevents air being trapped – poor insulator
	Has a light shiny lustre which prevents soiling
	The structure has cavities, making it highly absorbent
Silk	The only natural long, continuous filament
	Consists of two long protein bundles closely packed together
	Highly absorbent, making it very cool to wear
Wool	Fibres have a natural **crimp** (waviness) and are coated with scales
	The crimp allows air to be trapped, so it is a good insulator
	The scales can hook together when wet, causing shrinkage
	The natural grease in the fibres makes wool water repellent
Polyester	Can be engineered for different purposes – versatile
	A long flat filament that does not hold water – poor absorbency
	Can be crimped in the manufacturing process, allowing it to trap some air and improve insulation

Spinning

- Spinning is the process of twisting fibres together to make a yarn. The individual fibres are quite weak but gain additional properties when they are spun into yarns.

Now test yourself answers at www.hoddereducation.co.uk/myrevisionnotes

- There are two ways in which fibres are spun:
  - S twist (anticlockwise)
  - Z twist (clockwise).
- A tight twist squeezes out air, making the fibres closer together, resulting in a yarn that is strong but not warm. A loose twist allows more air to be trapped, making the yarn warmer but much weaker.

Fancy yarns

- Twisting multiple yarns together creates textured or novelty yarns with irregular surfaces and varying thicknesses such as bouclé.
- These add texture and surface interest when knitted or woven.

Quilting

- English quilting consists of three layers: a top layer, a plain lower layer and a layer of polyester wadding sandwiched between them.
- Reasons for quilting include:
  - insulation – air is trapped in between the layers, which will keep the wearer warm
  - decoration – adding surface interests
  - functional reasons – **reinforcement** where added protection is needed.

Components

- Almost all fashion and textiles products require components to be able to function.
- They are also used for aesthetic reasons.
- For more on components, see Section 3, Topic 7.

S twist

Z twist

Figure 3.12 Spinning yarns – S twist, Z twist

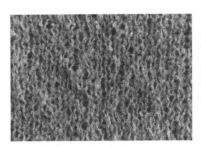

Figure 3.13 Knitting with bouclé yarn creates an interesting textured fabric

> **Reinforcement:** extra material added to increase strength.

Exam tip

The central cavity within cotton and linen fibres enable these fibres to soak up and store moisture. When discussing the properties of cotton, it is important to note that the fibres are absorbent. The ability to wick away moisture makes them appear breathable and cool to wear. Absorbency is the important property.

Typical mistake

Key terms relating to the construction of fabric are often confused or misunderstood. Students typically mix up warp and weft yarns: warp yarns run vertically along the length of fabric following the straight grain while weft yarns run horizontally across the fabric, interlinking with the warp yarns.

6 Selection of materials and components

Choosing material and components

REVISED

There are a number of factors to consider when choosing materials and components for fashion and textiles products:

- Aesthetic qualities: consumers will choose products based on appearance, so the colour and pattern must be suitable. Texture and lustre must also be appealing to customers and the weight of the fabric must also be suitable.

- Physical qualities: the weave used (see Section 3, Topic 2) will affect the way in which the fabric will handle. The density and the drape of the fabric will also affect the product.
- Economic: fabrics such as silk are more expensive and using these will increase the cost of the finished product.
- Performance: the construction of the fabric, properties of the fibre and finish to be used all need to be considered to make sure the final product performs as intended.

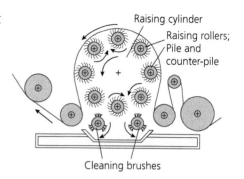

Figure 3.14 **Wire brushes raise the fibres to produce a soft fluffy surface**

Table 3.9 **Mechanical and physical finishes**

Process	Aesthetic and functional effect
Brushing	Wire brushes are passed over the surface of the fabric to raise the fibres This produces a soft, fluffy surface, improving its appearance and enhancing insulation
Calendering	Fabric is passed through heated rollers to give it a smooth, more lustrous surface, enhancing its aesthetic qualities Moiré is a variation of this finish and gives a wavy watered effect
Glazing	This is a similar process to calendering but with stiffeners or resins added to the finish to make it more permanent
Embossing	Heat and pressure are applied to fabric as it passes through engraved rollers; this leaves an imprint or slightly raised surface pattern in the fabric

Table 3.10 **Chemical finishes**

Process	Aesthetic and functional effect
Mercerising	Caustic soda is used to make the fibres in the fabric swell, leaving a more lustrous, stronger fabric This process works for cellulose fibres only
Crease resistance	A resin coating is applied to the fabric, stiffening the fibres and making products easier to care for Treated fabrics dry more quickly but reduce the ability to absorb moisture
Flame resistance	Chemicals such as Proban® are applied to the surface of fabrics as a liquid coating, which reduces the fabrics' ability to ignite and burn
Bleaching	Bleaching removes any natural colour and is used to prepare fabric for dyeing and printing
Stain resistance	Teflon™ and Scotchguard™ are fabric protectors mostly used on clothing and home furnishings
Anti-static	A chemical-based product is applied to the fabric to prevent the build-up of electrostatic charge, also known as static electricity
Water repellence	Silicon is applied as a semi-permanent finish which repels water Applying a fluorochemical resin makes fabric water repellent and wind resistant Teflon™ and Scotchguard™ are water-repellent finishes Coating with PVC, PVA or wax also repels moisture
Shrink resistance	A chlorine-based chemical smooths out the scales on the wool fibre which prevent them locking together during washing, preventing shrinkage
Moth proofing	This finish repels moths which feed off the keratin found in wool fibre

Table 3.11 **Biological finishes**

Process	Aesthetic and functional effect
Stone washing	Stones added to industrial washing machines give a distressed 'worn-out' look to fabrics. Often applied to denim jeans

Responsibilities of designers and manufacturers

Designers and manufacturers should consider the following:

- Most applied **finishes** use energy and water as well as chemicals and toxins which are hazardous to the environment and the health of the textile workers.
- Chemical finishes applied to fabrics reduce the fibre's ability to fully degrade; chemical traces damage the delicate ecosystems and impact on biodiversity.
- New finishing chemicals are being developed that are reusable, do not require the use of water and are biodegradable.

> **Finishes:** added to fabrics to improve their aesthetics, comfort or function. These finishes can be applied mechanically, chemically or biologically.

> **Exam tip**
>
> To demonstrate knowledge and understanding, know how finishes are applied to fabrics and be able to explain the purpose of the finish, particularly in relation to the end product.

> **Typical mistake**
>
> Quite often answers relating to environmental issues are not fully explained. You will not gain marks by simply stating something is bad for or has a negative impact on the environment. Always try to give a specific example in your response and a full explanation of the impact.

7 Stock forms, types and sizes

Stock forms

REVISED

- Standard widths for textile fabrics include 90 cm (interfacings or linings), 115 cm, 150 cm, 200 cm.
- Cotton sheeting at 240 cm wide is also available from some specialist fabric retailers.
- 'Fat quarters' are pre-cut fabric pieces measuring 45 cm × 55 cm, ideal for small projects.

Table 3.12 **Common fabric names**

Fabric name	Example of use	Fabric name	Example of use
Denim	Jeans, dresses	Cotton poplin	Blouses, dresses
Voile	Curtains	Jersey	Knitted T-shirts, leisure wear
Chiffon	Evening wear, lingerie	Felt	Crafts, hats, snooker tables
Corduroy	Trousers, skirts, jackets	Tweed	Jackets, coats
Velvet	Eveningwear, jackets	Drill	Heavy-duty workwear
Lace	Decorative, clothing	Organza	Bridal, high fashion

Table 3.13 **Standard components**

Component	Types available include:
Threads Types of thread are available in different weights, strengths and textures for different uses	Machine thread Tacking thread Top stitching Embroidery thread Monofilament
Fastenings These are selected according to their function and intended purpose as well as aesthetic qualities	Buttons Poppers Zips Velcro® Hooks and eyes Toggles
Structural components These are used to add support and help shape a product	Boning Petersham
Other components These components provide functional support or are chosen for aesthetic reasons	Bias binding Elastic Ribbons Laces Beads Sequins Eyelets

Figure 3.15 **The organza overlay fabric is sheer and lightweight**

Figure 3.16 **An assortment of textile components that support textile products**

Cost and quantities

The cost of raw materials and components should be taken into consideration when making a textile product. The following are points to consider:

- the cost of the fabric and the width that will give the best use of fabric
- the fibre type, which can affect the overall price
- all component parts need to be costed
- bulk buying of fabric and components in industry can substantially reduce the overall cost of raw materials
- pattern templates must be laid correctly to minimise the amount of waste fabric.

Typical mistake

In questions where calculations are needed, such as working out costs, make sure you show all of your workings even if you have used a calculator. You will lose marks if you do not show the method used.

Exam tip

Know the common names of textile fabrics such as corduroy, chiffon or cotton poplin and not just by their fibre source. In an examination it demonstrates a higher level of understanding. Cotton, for example, is the name of the fibre and there are many different variations when it is made into fabric.

8 Manufacturing to different scales of production

Scales of production

REVISED

- Scales of production depend on the quantity needed, the complexity of products, timescale and budget.
- Scale of production affects the quality and cost of products.

One-off, bespoke or job production

- **One-off**, custom-made or **bespoke** products are made by an individual or a small team of highly skilled, versatile workers with the ability to adapt to a range of processes and machinery.
- One-off products are often commissioned as unique pieces by clients who may also have direct input into the design process.
- They often use high-quality fabrics and components and are therefore expensive to purchase.
- Many haute couture gowns are made in this way.

> **One-off**: a single product.
>
> **Bespoke**: made to measure for an individual client.

Figure 3.17 **Bespoke wedding dresses are made for individual clients**

Batch production

- **Batch production** is used to produce a specific number of identical products in a set timescale, such as seasonal fashion products, and are usually of mid to low quality.
- The products are made by large teams of workers who specialise in one element of the construction process, which means work can be repetitive and boring.
- Batch production is flexible and can change to meet market demand. Repeat orders can be facilitated quite easily.
- Ready-to-wear (prêt-à-porter) designer collections are made using this method as a small number of well-made products are required.

> **Batch production**: a number of identical products are produced.

Mass production

- **Mass production** is the largest scale of production, usually for products that are in high demand over a long timescale. Many factories run 24/7 to maximise output and profit.
- Typical mass-produced products include socks and plain T-shirts where styles rarely change.
- Workers are skilled or specialised in one element of the construction process.
- CAM is increasingly used in mass production where consistency and speed are important.

> **Mass production:** large quantities of identical products are produced.

Manufacturing systems

Mass and batch production lines can be organised in different formats in order to maximise efficiency and output:

- Straight line production: the work is passed from one worker to the next, either along a conveyor belt or on overhead automated systems.
- Progressive bundle: bundles of garments or product parts are moved in sequence from one worker to the next. Each worker completes a single operation within the bundle before passing it to the next worker.
- Cell production: groups of workers operate together to make whole or part products. Cells operate separately within other systems.

> **Typical mistake**
>
> You may lose marks if descriptions between mass and batch production do not differentiate sufficiently between the two systems. Similar products can be made under both systems but quantities and timescales will be considerably different.

The role of designers

REVISED

- Fashion designers create new styles targeting different parts of the market.
- They often create 'statement pieces' which other designers may take inspiration from to create their own more commercially viable products, with cheaper fabrics and components.
- The essence of the original look is maintained to keep the product on-trend.
- Designers are influenced by a number of factors, including the world around them, fashion forecasters, street styles and developments in the main fashion centres such as London, New York, Paris, Tokyo and Milan.

Influences on fashion

- Media – television, films, magazines that feature street and music trends, celebrities.
- Lifestyle choices – leisure activities and sports clothing, travel and inspiration from traditional clothing from other countries.
- History – the history of fashion over the centuries.
- New technology – advances in technology can bring about new options for designers, e.g. Gore-Tex was developed for use in space.

Trendsetters

- Trendsetters have a strong sense of style and fashion and their choice of clothes make an impact.
- Celebrities may be seen as trendsetters, particularly if they become identified with a particular style.

Now test yourself answers at www.hoddereducation.co.uk/myrevisionnotes

Image makers

- Image makers specialise in putting together a particular 'look' or style for their clients, who could be individuals, businesses looking for a corporate image, or particular products.
- They understand **contemporary** trends and their clients' needs and use these to create a look that will give a positive image of the client.

Fashion forecasting

- Fashion forecasters predict trends and produce reports for designers and manufacturers to help them to design new products.
- Their predictions are based on research and analysis of trends in fabrics, colours, details and features.
- Predictions are made well in advance so forecasters need to be able to identify upcoming trends.

Contemporary: fashion and styles that are currently popular.

Exam tip

When a question starts with 'Explain', you must state a fact and provide further elaboration of that fact. You will lose marks if you simply list different facts.

9 Specialist techniques and processes

Tools and equipment

REVISED

Hand tools for manufacturing textile products include:

- metre rule: for marking out and cutting fabric, and measuring hemlines
- tape measure: for taking body measurements
- craft knife: for cutting small templates and stencils
- cutting mat: includes guides for accurate cutting and used with a rotary cutter or craft knife
- quick unpicker: to unpick faulty seams
- pins: a temporary method of holding fabric together.

Sewing machines

- Domestic sewing machines have a wide range of facilities and stitches to complete different processes.
- Computerised sewing machines can also embroider original designs.
- Domestic machines have optional feet or specialist attachments for specific processes such as attaching zips.
- Sewing machine needles vary and should be selected according to the fabrics being used.
- An overlock machine is used in industry to cut a straight edge on the material and oversew the edge to neaten the seam in one process.

Laser cutting

- Laser cutting is controlled by a CAD program.
- The design is drawn as a 2D image which the laser follows to accurately cut or etch the design.
- The laser strength and speed are set depending on the material to be cut.
- Laser cutters are used in industry to cut through multiple layers of fabric on automated systems.

- Disadvantages include:
  - not all fabrics can be cut as some melt and burn
  - the laser leaves an unsightly burnt brown edge on some fabrics.

Pattern markings and cutting

- Pattern templates include pattern markings which must be followed to ensure that the finished product is the correct size, shape and quality.
- The templates are placed on fabric and then cut around to create the pieces of fabric for the product.
- Fabrics can be laid out in different ways:
  - lengthways and folded over so that selvedge edges come together; the templates are then laid out according to pattern instructions
  - crosswise fold (folded along the width of the fabric) for a more efficient lay plan
  - on the bias, at a 45-degree angle to the straight grain, creating fabrics that are more flexible and can easily be sewn into curved shapes but are wasteful.

Table 3.14 **Pattern markings that must be followed accurately when making textile products**

Pattern mark	Meaning of the mark	Why it is important
←——————→	Straight grain or grain lines	Template must be parallel to the selvedge edge, so the garment hangs as intended or lies flat
↓——————↓	Place on folded edge	The edge of the template needs to be on a folded edge of the fabric, as the piece is symmetrical
———————	Adjustment lines to lengthen and shorten templates	The templates can be adjusted here to get a more personalised fit
···············	Cutting lines in various sizes	Cut along the desired size
– · – · – · –	Stitching line	This is where stitches should be when joining pieces – if not adhered to, the product will not fit together
——✂——	Seam allowance	The distance of the stitching line to the fabric edge, usually 1.5 cm
●	Dots	Indicates the position of a component or shaping technique
◆ ◆◆	Notch	Indicate how pieces fit together and how to match a pattern, such as stripes
⊙	Position of button	Transfer the mark onto the fabric for correct placement on the garment
⊢——⊣	Position of button holes	Transfer the mark onto the fabric for correct placement on the garment
⊰	Position of dart	The dots need to match up to create the dart
⊏≢⊐ ⊏≢⊐	Placement of pleats and tucks	The lines need to match up to create the pleat or tuck

Tools used for transferring pattern markings to fabric include tailor's chalk, vanishing markers, tracing wheel and carbon transfer, tailor's tacks and hot notch markers.

Cutting tools

Tools for cutting fabric include:

- fabric shears, which have long blades to easily cut fabric
- embroidery scissors, which have a sharp pointed blade for cutting intricate work
- pinking shears to produce a cut zig-zag shape along the raw edge of a seam, which prevents fraying
- rotary cutters, which are sharp and accurate tools that are rolled along the surface of the fabric
- laser cutters to cut complicated pattern pieces
- band saws, which are used in industry to cut through large numbers of layers of fabric accurately and quickly
- automated die cutters, which are used for cutting constant shapes through several layers of fabric.

Figure 3.18 Tailor's tacks

Figure 3.19 Fabric shears are the most commonly used method for pattern cutting at home or at school

Seam construction

REVISED

It is important to use the correct type of seam when joining textile fabrics. The method used will depend on the fabric and product.

Tolerance and allowances

- The correct **seam allowances** and **tolerances** must be used throughout the manufacture of a product.
- The standard seam allowance used in textiles is 1.5 cm.
- If the correct seam allowance is not used consistently then product parts will not fit together as intended, leading to a poorer quality product.
- Some textile products are more complex so a permitted tolerance of a seam of about +/–1 cm is acceptable. However, this could still affect the overall size.

> **Seam allowance:** the distance between the raw edge of the fabric and the stitching line.
>
> **Tolerance:** an allowance included in the seam allowance for inconsistency when assembling a product.
>
> **Raw edges:** fabric edges that are not neatened – unfinished.

Types of seam

- Plain seams are the most common seam type. Seam allowances can be stitched together or pressed flat open.
- Double-stitched seams are the same as plain seams but with an additional row of stitching about 5 mm away from the first row, which creates a stronger seam.
- French seams are enclosed and hide all **raw edges**. They are used on more expensive clothing and sheer materials where seams need to be hidden.
- Flat-felled seams are strong seams with two rows of stitching adding to strength. The top stitching is often in a different colour.
- Lapped seams are created by overlapping one layer on top of the other before stitching. This is similar to flat-felled seams.
- Clipping seams occur when a curved seam needs to be clipped along the seam allowance to allow fabrics to lie flat, particularly around rounded areas such as necklines and armholes.

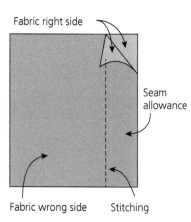

Figure 3.20 **Plain seam**

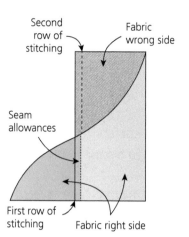

Figure 3.21 **French seam**

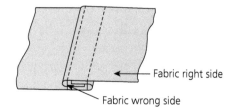

Figure 3.22 **Flat-felled seam**

Figure 3.23 **Seam finished with a zig-zag**

Figure 3.24 **Edge finished with a bias binding**

Finishing seams

Seams are finished to prevent fraying. Methods include:

- overlocking: excess fabric along the seam is trimmed off, the stitch overlaps the edge
- zig-zag stitch: sewn along the raw edge of a seam
- pinking shears: cut a zig-zag finish and prevent fraying
- seams and fabric edges can be bound using a bias binding.

Table 3.15 **Methods for adding body and shape**

Method	Technique
Pleats	Fabric is folded back on itself and sewn in place, narrowing the original width of the material but adding shape
	Can be used as a decorative frill on products
Tucks	Similar to pleats and draw in fullness
Gathers	The fabric edge is gently drawn in to reduce and narrow the original width of the fabric, giving fullness and shape to a product
	Gathers can also be used as a decorative feature
Darts	Darts are made by creating folds in the fabric that taper to a point to improve shape and fit
	Darts are commonly used around the bust area but can be used anywhere to give shape
Princess line seams	Used to create a tight-fitting garment that follows the contours of the body
	Darts are joined from mid-armhole through the fullest part of the bust, down to the waistline and hip to the hem
	Often found on tops and swimwear

Figure 3.25 **The pleats are stitched close to the waistband and lie flat but give shape to the skirt**

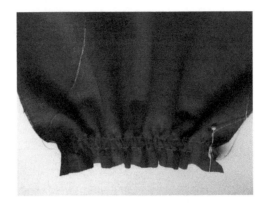

Figure 3.26 **Gathers are formed by sewing two rows of stitching and pulling up the thread ends to create gathers**

Edge finishes

- Finishes applied to the edge of textile fabric depend on the fabric and product.
- Hems vary depending on the fabric and product, for example in some cases stitches may be visible while for others it will need to be hidden.
- Piping can be inserted into seams to add structure and help make a product more durable.
- Other types of edge finish include frills and binding.

A basic double-folded hemline with a machine stitch

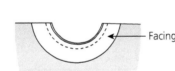

The facing is stitched in place and the seam is then clipped before the facing is turned to the inside

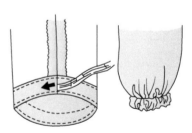

The elastic is inserted into a casing to form a cuff

Figure 3.27 **Different methods of finishing edges**

Style details

- Style details add shape and form to a product.
- Some of the most commonly used details are shown in Figure 3.28.

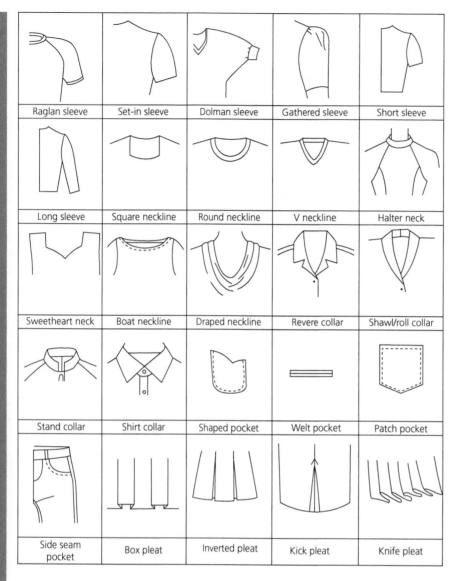

Figure 3.28 Style details

Computer-aided design and manufacture

REVISED

Computer-aided design (CAD)

- Designers use **CAD** to develop surface designs for printing that can be transferred to relevant digital printing systems.
- Designs can be revisited, manipulated and changed, including the development of patterns and colourways.
- CAD is a more cost-effective way of developing designs. By using 3D imagery or virtual prototyping, the need to make numerous prototypes is reduced. It is considered a more sustainable way of designing.
- CAD software is used to develop digital **lay plans**, allowing manufacturers to **tessellate** the pattern pieces to maximise fabric usage.

Computer-aided manufacture (CAM)

- Computers can be used to control sections of manufacturing in the textile industry. Some machinery is semi-automated, requiring some human input, while other machines are fully automated.

CAD: computer-aided design.

Lay plan: how templates are laid out on fabric.

Tessellate: how pattern templates fit together to use the least amount of fabric.

- **CAM** systems are expensive but speed up manufacturing, improve productivity and consistency and reduce human error.
- CAM applications include multi-head embroidery machines, digital printing on fabrics, laser cutters, 3D printers and automated fabric spreaders.

10 Surface treatments and finishes

Surface decoration techniques, finishes and treatments, also known as **embellishments**, are applied to textile fabrics and products for aesthetic reasons, for example colour and texture.

CAM: computer-aided manufacture.

Surface decoration techniques: used to improve the aesthetics of a product by adding colour, texture and pattern, for example dyeing, printing and embroidery.

Embellishments: surface decoration such as embroidery and beading.

Dyeing

REVISED

- Dyeing is the most common method of colouring fibres and fabrics.
- Natural dyes work well on natural fibres, while synthetic dyes will give deeper or brighter colours.
- Synthetic fibres need chemicals to take on the dyes.

Methods of dyeing textile fabrics include:

- Dip dyeing: this produces a graduated colour effect. Part of the fabric is dipped in the dye and is then gradually removed from the dye bath.
- Random: this involves dyeing or colouring small sections of fabric or yarns; different sections have different colours with no regularity to the design.
- **Piece dying**: an entire length of fabric is dyed one colour.
- Tie and dye: the fabric is tied or knotted in a variety of ways to produce unique and varied designs. This is called a **resist method** of colouring.
- Batik: this is also a resist method where hot melted wax is applied to fabric in the desired pattern. Once cooled, the fabric is submerged in a dye bath or the dye can be directly painted on.

Piece dyeing: an entire length of fabric is dyed one colour.

Resist method: a means of preventing dye or paint from penetrating an area on the fabric. This creates the patterns.

Printing

REVISED

Types of printing include:

- **Flatbed screen printing**: screens are used to apply a different colour and design to the fabric as it moves along a conveyor belt. When the design is complete the fabric will be fixed, washed and pressed.
- **Rotary screen/roller printing**: this is similar to flatbed printing except cylinders are used instead of screens. The fabric passes underneath the spinning cylinders, printing a continuous pattern onto the surface of the fabric. This is also referred to as roller printing.

- **Silk screen printing**: a fine mesh fabric is stretched over a wooden frame with part of the screen masked out with an opaque paste. The screen is laid face down on the fabric, printing ink is placed on the underside of the frame and a **squeegee** is used to drag the ink across the screen, forcing the ink through the fabric to leave the design on the material.

- Stencilling: a stencil is usually made from a thin sheet of card or plastic with a pattern cut out of it. Paint or dye is then applied through the holes in the stencil to leave a design on the fabric. Stencils can be hand cut or on the laser cutter.

- Block printing: this is a traditional method of printing in which paint or dye is applied to a relief block which is then pressed onto fabric to create a pattern. The process is repeated to create an overall repeat design.

- Digital fabric printing: large inkjet printers and specialist dyes transfer a digital image to the surface of the fabric. The technique can be used to create intricate and detailed images and to test sample pieces before production.

- Discharge printing: a bleaching agent is applied to fabric, in the required design. The bleaching agent destroys the colour, leaving a white or pale design.

Painting REVISED

- Fabric paints, fabric felt pens and pastel crayons are applied directly onto textile fabrics to create the desired design. Dimensional paints are also applied directly to fabric to give a raised, slightly 3D surface.

- Specialist silk paints give a very delicate watery effect and can be used with Gutta outliner, which acts as a barrier to separate sections of the design.

Transfer printing

- Sublimation printing uses heat and pressure to transfer dye from specialist printer paper, turning the dye into a gas that binds directly to the fibres, leaving a crisp design.

- An inkjet printer can be used in a similar way, using specialist transfer paper to transfer a design to fabric to be fixed in place either with a heat press or an iron.

Figure 3.29 **A hand-painted silk scarf which uses the Gutta outline to create a barrier between the different sections**

Embroidery REVISED

Embroidery designs vary depending on whether they are prepared by hand or by machine. Types of machine embroidery include:

- Free machine embroidery: the fabric is secured in a frame and the machinist moves the fabric around freely to create a design.

- Machine embroidery: in-built decorative stitches can be used to enhance any designs. Computerised sewing machines either have pre-installed designs or are linked to CAD packages to create and then stitch original designs (CAM).

- Appliqué: this is a traditional way of applying a design to fabric by stitching separate pieces of fabrics onto a base fabric. There are endless possibilities for creative design work through mixing colour, textures, stitch type and number of pieces.

Figure 3.30 **A modern interpretation of an appliqué design**

Embroidery is often further enhanced with beads and sequins.

- Beadwork: beads are used to enhance other techniques such as appliqué and can be placed randomly, used to outline an area or in a cluster. Beads are available in many different shapes, sizes and material. Similar effects can be achieved using sequins or diamantes.
- Patchwork: small pieces of cloth of different designs, colours or textures are sewn together, either in regular repeat patterns or in more abstract designs. This is an effective way of reusing old clothing.

Laser engraving

See laser cutting in Section 3, Topic 9.

> **Exam tip**
>
> 'Evaluate', when used as a command word in exam questions, requires evidence of appraisal or making judgements. Statements made should reference both positive and negative viewpoints.

> **Typical mistake**
>
> Show understanding of the stages needed to apply surface treatments and finishes to fabrics. Stages in a process need detail to fully explain the stage. Marks may not be awarded for a sequence that lacks detail, is incomplete or is in an incorrect order.

Now test yourself

TESTED ☐

1. Explain why a satin weave fabric is less stable than a twill weave fabric. [3]
2. Give three reasons why the intense farming of cotton crops is detrimental to the ecosystem. [3]
3. Explain the importance of following pattern marking when laying out templates on fabric. [2]
4. Explain the purpose of a cell operating within a manufacturing system. [3]
5. Outline the reasons for using different seam construction methods in product manufacture. [4]

Exam practice

1. Weaving and knitting are the two main methods of fabric construction.
   (a) Label the diagram below to show where the listed features would be seen on a length of woven fabric. [4]
   selvedge edge   warp yarn   bias line   weft yarn

   (b) Describe two reasons for weaving fabric in different ways. [2]
   (c) Explain why knitted fabrics are often used in casual leisure clothing. [3]
2. Fibres are the raw materials of textiles.
   (a) Underline the two natural fibres in the list below. [2]
   silk   polyester   acrylic   linen

(b) Explain how the structure of the cotton fibre allows it to absorb moisture. [2]

(c) Evaluate the use of fibre blends and mixes when choosing fabrics for textile products. [5]

3 The textile industry has a major impact on the environment.

(a) Explain why synthetic fibres have a negative impact on the environment. [3]

(b) Describe the impact that cotton fibre crops have on the environment. [4]

(c) Explain how the transportation of textile goods affects our carbon footprint. [3]

(d) Outline the impact a throwaway culture has on:

(i) workers in developing countries [2]

(ii) the environment. [2]

4 Finishes are applied to textile fabrics for different reasons.

(a) Explain the purpose of a Scotchguard™ finish. [2]

(b) Explain how brushing enhances the functionality of cotton fabrics. [3]

5 Components are used in textile products for a range of different reasons.

(a) Give two reasons for using piping as an edge finish on a textile product. [2]

(b) A textile student has decided to make a circular floor cushion that will have a piped edge inserted into the seams of the two circular end pieces, as shown on the diagram below.

The diameter of each circular end piece is 70 cm.

The height of the floor cushion is 30 cm.

Diameter: 70 cm

Piped edge

Height: 30 cm

(i) Calculate how much piping cord will be needed to complete the cushion. [4]

(ii) The sides of the floor cushion will be cut from one rectangular piece of fabric. Calculate the size of the rectangular template that will be needed to form the side of the floor cushion. Include seam allowances of 1.5 cm in your calculation. [2]

6 Fashionable clothing is made using different scales of production.

(a) Explain why batch production would be a suitable scale for the manufacture of children's woollen coats. [3]

(b) Describe the benefits to the client of having a bespoke product made. [4]

7 Pattern language is important in the manufacture of textile products.

(a) State the name of the following pattern markings as seen on pattern templates. [3]

←——→	—◆—	══
....	....	....

(b) Describe the purpose of a tailor's tack. [2]

(c) Analyse the use of CAD in lay planning in garment production. [5]

8 Decorative techniques are used throughout fashion and textiles.

(a) Place a tick (✓) to show which of these statements are true. [5]

- Dip dyeing fabric gives a graduated effect to the colour.
- Marbling is a resist method of adding colour to fabric.
- Roller printing is another name for flatbed printing.
- Free machine embroidery allows the fabric to move freely while being stitched.
- Batik involves using hot melted wax to outline a shape on fabric.

(b) Computer-aided design is used throughout the textile industry. Evaluate the use of CAD in the design and development of digital printing on textile fabrics. [6]

ONLINE

4 Product design

1 Paper and boards

Paper

REVISED

- Paper and board are available in standard sized sheets.
- Sheets range from A10 (approximately the size of a postage stamp) through to 4A0 (larger than a king-size bed sheet).
- The most common sizes used by designers are between A6 and A0.
- Each sheet size is twice the size of the one before, for example A3 is twice the size of A4.
- If we fold a sheet of paper in half it then becomes the next size below, for example an A1 sheet folded in half becomes A2 size.

Size	A10	A9	A8	A7	A6	A5	A4	A3	A2	A1	A0	2A0	4A0
Length (mm)	37	52	74	105	148	210	297	420	594	841	1189	1682	2378
Width (mm)	26	37	52	74	105	148	210	297	420	594	841	1189	1682

most common sizes used by designers

Figure 4.1 **Paper sizes**

- The thickness of paper is known as its 'weight'.
- Weight is measured in grams per square metre (g/m^2 or **gsm**).
- A weight greater than 170 gsm is classified as a board rather than a paper.
- Boards are usually classified by thickness as well as by weight.
- The thickness of board is measured in **microns** (one-thousandth of a millimetre).

> **Gsm**: grams per square metre – used to measure the weight of paper.
>
> **Micron**: one thousandth of a millimetre (0.001mm) – used to specify the thickness of card.

Table 4.1 **Common paper types**

Paper type	Properties	Uses
Cartridge paper	Available in different weights between 80–140 gsm More expensive than layout and copier paper Has a slightly textured surface and is slightly creamier in colour Ideal surface for pencil, crayons, pastels, water colour paints, inks and gouache	Sketching, drawing and painting
Copier paper	Weighs approximately 80 gsm Smooth surface	Printing, photocopying and general office purposes
Bleedproof paper	Available in similar thicknesses to cartridge paper Smooth surface Bleached bright white Stops marker 'bleed'	Drawing and sketching using marker pens

Table 4.2 **Card and cardboard**

Board type	Properties	Uses
Card	Thin card is around 180–300gsm in weight. Available in a wide range of colours, sizes and finishes. Easy to fold, cut and print on	Greetings cards, paperback book covers and simple models
Cardboard	Around 300 microns upwards in thickness. Inexpensive and can be cut, folded and printed onto easily	Packaging, e.g. cereal boxes, tissue boxes, sandwich packets. Modelling design ideas and making templates for parts made from metal or other resistant materials
Mounting board	A rigid type of card of around 1.4mm thickness (1400 microns). Available in different colours but white and black are the most common	Picture framing mounts and architectural modelling
Foam board	Lightweight board made of polystyrene foam sandwiched between two pieces of thin card or paper. Smooth surface. Lightweight and rigid. Available in a range of colours and thicknesses	Modelling. Point-of-sale displays
Solid white board	High-quality cardboard made from top-quality bleached wood pulp. Suitable for highly detailed printing as it gives a clear, sharp image	Hardback books. Packaging for expensive perfumes and make-up
Corrugated cardboard	A strong but lightweight type of card. Made from two layers of card with another, fluted sheet in between. Available in thicknesses ranging from 3mm (3000 microns) upwards	Used for packaging fragile or delicate items that need protection during transportation. Widely used as packaging for takeaway food due to good heat-insulating properties
Duplex board	Made up of two layers that are mostly made from waste paper pulp. Lightweight with high strength properties. Usually has a smooth, white, medium gloss finish but also available in a wide variety of other finishes	Packaging, particularly for food and drinks cartons, clothing and pharmaceutical goods

Paper finishes

- **Coatings** can be applied to paper to improve the **opacity**, lightness, surface smoothness, lustre and colour-absorption ability.
- Cast coatings are when the wet paper is coated with china clay, chalk, starch, latex and other chemicals, then rolled against a polished, hot, metal drum, creating a smooth, reflective and shiny surface that produces sharper, brighter images when printed on.

> **Coating:** an additional outer layer added to a product.
>
> **Opacity:** lacking transparency or translucence.

Super calendering

- Super calendering is when paper passes through a calender or super calender.
- This is a series of rollers with alternately hard and soft surfaces that press the paper to create a smoother and thinner paper with a very high-lustre surface.
- Super calendered paper is primarily used for glossy magazines and high-quality colour printing.

Table 4.3 **Types of paper finish**

Type of paper	Properties	Used for:
Cast coated paper	Provides the highest gloss surface of all coated papers and boards	Labels, covers, cartons and cards
Lightweight coated	A thin, coated paper, which can be as light as $40\,g/m^2$.	Magazines, brochures and catalogues
Silk or silk matt finished papers	Smooth, matt surface. High readability and high image quality	Product booklets and brochures
Calendered or glossy paper	Glazed shiny surface – can be coated and/or uncoated	Colour printing
Machine-finished paper	Smooth on both sides. No additional coatings applied after leaving the papermaking machine	Booklets and brochures
Machine coated	Coating applied while it is still on the paper machine	All types of coloured print
Matt-finished paper	Slightly rough surface prevents light from being reflected. Can be both coated and uncoated	Art prints and other high-quality print work

Card and board finishes

REVISED

Varnish

- Varnish coatings enhance the look and feel of graphic products.
- Spirit varnish is used to give a high-shine finish that feels like plastic to the touch.
- Spot UV varnish is a special varnish that is cured or hardened by **UV light** during the printing process. Spot varnish is used on certain areas of the paper or card only to make it shinier and to stand out.

> Ultraviolet (UV) light: outside the human visible spectrum at its violet end.

Hot foil application

- Hot foil application is used to produce metallic finishes such as gold or silver.
- It is often used for lettering on invitations and business cards.

2 Natural and manufactured timber

Hardwoods

REVISED

- Hardwoods come from deciduous trees.
- Most deciduous trees lose their leaves in autumn.

- Deciduous trees take a long time to mature.
- Hardwoods generally have a close grain structure. This makes them harder and stronger than softwoods, although this is not always the case.
- They can be sanded to a finer, smoother finish and be given a higher-quality finish.
- Hardwoods are generally more expensive than softwoods.

> **Typical mistake**
>
> When asked to describe the life cycle of a product that is made from natural timber, students will often start with the finished product and concentrate on how it is disposed of. You must start from its primary source, the tree.

Table 4.4 **Hardwoods**

Hardwood	Properties	Common uses
Jelutong	A close-grained timber with a pale colour Medium hardness and toughness Easily worked	Pattern making
Beech	A hard, strong, close-grained timber with a light brown colour with distinctive flecks of brown Prone to warping and splitting Can be difficult to work	Furniture, children's toys, workshop tool handles and bench tops
Mahogany	A strong and durable timber with a deep reddish colour Available in wide planks Fairly easy to work but can have interlocking grain	Good-quality furniture, panelling and veneers
Oak	A hard, tough, durable, open-grained timber Can be finished to a high standard	Timber-framed buildings, high-quality furniture, flooring
Balsa	A very lightweight, soft and easily worked timber Pale in colour Weak and not very durable	Model-making, floats and rafts

Softwoods

REVISED

- Softwoods come from coniferous trees, also known as evergreens.
- Coniferous trees are quick growing and take around ten years to reach maturity.
- Most softwoods have an open grain and are generally less dense and not as strong as hardwoods.

Table 4.5 **Softwoods**

Softwood	Properties	Common uses
Western red cedar	Very resistant to weathering and decay Has a light reddish grown colour with a close, straight grain Easily worked	Fencing, fence posts and cladding
Scots pine	A straight-grained, light yellow-coloured timber Soft and easy to work Can be quite knotty	Interior joinery and furniture, window frames
Parana pine	Has a distinctive open, straight grain Contains few knots and is strong and durable	Interior joinery and staircases

Now test yourself answers at **www.hoddereducation.co.uk/myrevisionnotes**

Manufactured boards

- **Manufactured boards** were developed as an alternative to natural timbers.
- They fall into two categories: laminated boards and compressed boards.
- Laminated boards are produced by gluing together large sheets or **veneers**.
- Compressed boards are manufactured by gluing together particles, chips or flakes under pressure. They are often covered with thin plastic laminate.
- Manufactured boards are available in large sheets. They are generally less expensive than natural timbers.

> **Manufactured boards:** sheets of timber that have been manufactured to give certain properties.
>
> **Veneers:** thin sheets of natural timber.

Table 4.6 **Examples of manufactured boards**

	Properties	Uses
Laminated boards, e.g. plywood, blockboard, veneered board	Strong	Shelving
	Look like 'real' wood	Workbenches and worktops
Compressed boards, e.g. medium-density fibreboard (MDF), chipboard, hardboard	Smooth surface	Kitchen worktops
	Can be coated with plastic laminate	Cupboards
	Easy to cut and shape	Bedroom furniture
	No grain	

Finishes

REVISED

- Finishes help protect timber from damage and enhance its appearance.
- Wood stains change the colour of the timber but don't add protection.
- Paint changes the colour of the wood and helps protect it from the weather.
- Varnish is a clear coating that gives protection against weather and enhances the appearance.
- Oils such as Danish oil and teak oil give timber an improved appearance and a low level of protection.
- Wood preservatives used on outdoor products protect from weather and help prevent decay.
- Manufactured boards can be varnished, stained and painted, but a coating must first be applied to seal the porous surface of some types of board such as MDF.

3 Ferrous and non-ferrous metals

- Metal is a naturally occurring material mined from the ground in the form of **ore**.
- The raw metal is extracted from the ore by crushing, smelting or heating.
- Metals are available in a variety of stock forms such as sheet, rod, bar, tube and angle.
- See Figure 2.23 (Section 2, Topic 8) for examples of stock metal forms.

> **Ore:** rock which contains metal.

Ferrous metals

- Ferrous metals are those that contain iron.
- Most ferrous metals are magnetic (due to the iron content).
- They are prone to corrosion when exposed to moisture and oxygen.
- To prevent rusting and enhance their appearance they can be painted, galvanised, plated and plastic coated.
- The properties of ferrous metals, such as hardness and malleability, are directly related to their carbon content – the more carbon that is found in steel, the harder and less malleable the steel becomes.
- Steel is the most widely used form of ferrous metal. It is easy to recycle.

Table 4.7 **Examples of ferrous metals**

Ferrous metals	Properties	Uses
Mild steel	Good tensile strength	Rolled steel joists (RSJs)
	Malleable (easy to cut and shape)	Car body panels, office furniture
Medium-carbon steel	High strength but less malleable	Gardening tools such as spades, trowels
High-carbon steel	Very high strength but even less malleable	Cutting tools such as saw blades and drill bits

Non-ferrous metals

- Non-ferrous metals do not contain iron and do not corrode.
- The majority are not magnetic, which makes them ideal for use in electronic devices and wiring.
- After steel, aluminium is the most widely used and recycled metal.
- Aluminium takes a huge amount of energy to extract and manufacture. It takes around 95 per cent less energy to recycle aluminium than to produce the raw material.

Table 4.8 **Examples of non-ferrous metals**

Non-ferrous metals	Properties	Uses
Aluminium	Extremely lightweight	Drinks cans
	Soft	Foils
	Malleable	Kitchen utensils
		Aeroplane parts
Copper	Good electrical conductivity	Plumbing pipes
	Good thermal conductivity	Electrical wire
	Corrosion resistance	Roofing
Brass	Alloy of copper, zinc and lead	Locks, gears, bearings
	Easy to cast	Ammunition casings
	Lustrous appearance	Musical instruments
	Low coefficient of friction	

Now test yourself answers at **www.hoddereducation.co.uk/myrevisionnotes**

Finishes

REVISED

- Ferrous metals require a finish to be applied to prevent them rusting.
- Ferrous metals can be painted, plated, galvanised, powder coated or polymer coated.
- Non-ferrous metals do not rust but will discolour due to oxidisation.
- Chrome plating, anodising and coating with clear lacquer can be used to finish non-ferrous metals.

4 Thermoforming and thermosetting polymers

Thermoforming polymers

REVISED

- Thermoforming polymers can be softened by heating and moulded into almost any shape.
- Once softened, or plasticised, they can be shaped and formed using a wide variety of processes, such as bending, **vacuum forming**, moulding and **extrusion**.
- Once the desired shape has been achieved, the polymer cools and maintains its new shape.
- Thermoforming polymers can be reheated, reshaped and cooled many times with minimal damage to the properties of the polymer.
- Thermoforming polymers can be recycled.

> **Vacuum forming**: a method of shaping a thermoforming polymer sheet by heating around a former.
>
> **Extrusion**: a length of polymer with a consistent cross-section.

Thermosetting polymers

REVISED

- Thermosetting polymers can be shaped and formed once only. They cannot be reheated or reformed once they have been formed and cooled.
- They make excellent insulators.
- They cannot be recycled.

Table 4.9 **Examples of thermoforming and thermosetting polymers**

	Type of polymer	Properties	Uses
Acrylic (PMMA)	Thermoforming	Hard	Car light units
		Excellent optical quality	Bath tubs
		Good resistance to weathering	Shop signage
		Scratches easily and can be brittle	Displays
		Excellent thermal and electrical insulator	
		Good plasticity when heated	
Polythene (PE)	Thermoforming	Tough	Carrier bags
		Flexible	Bin liners
		Easily moulded	Washing-up bottles

	Type of polymer	Properties	Uses
Polypropylene (PP)	Thermoforming	Semi-rigid Good chemical resistance Tough Good fatigue resistance Integral hinge property Good heat resistance	Buckets Bowls Crates Toys Bottle caps Car bumpers Jug kettles
Polycarbonate (PC)	Thermoforming	Tough, durable, impact resistant Good resistance to scratching Excellent thermal and electrical insulator Good plasticity when heated	Safety glasses Safety helmets
Expanded polystyrene (EPS)	Thermoforming	Lightweight Easy to mould Good impact resistance Excellent thermal insulator	Packaging Disposable cups and plates
Acrylonitrile butadiene styrene (ABS)	Thermoforming	Hard Excellent optical quality Good resistance to weathering Excellent thermal and electrical insulator Good plasticity when heated	Children's toys Phone casings Moulded-on electrical plugs
Polyvinyl chloride (PVC)	Thermoforming	Hard and tough Good chemical and weather resistance Low cost Can be rigid or flexible Excellent thermal and electrical insulator Good plasticity when heated Good tensile strength	Pipes, guttering, window frames
Nylon	Thermoforming	Hard, tough, resistant to wear A low coefficient of friction Excellent thermal and electrical insulator	Bearings Gears Curtain rail fittings Clothing
Urea formaldehyde (UF)	Thermosetting	Hardwearing Durable Non-conductor of electricity Waterproof Excellent thermal and electrical insulator	Electrical fittings Resin used in wood adhesives

	Type of polymer	Properties	Uses
Melamine formaldehyde (MF)	Thermosetting	Tasteless Odourless Shrink resistant Chemical resistant Excellent thermal and electrical insulator Stiff, hard and strong Scratch and impact resistant	Household crockery items (glasses, cups, bowls and plates) Toilet seats Pan knobs and handles Kitchen worktops
Carbon fibre-reinforced polymer (CFRP)	Thermosetting	Very strong Lightweight Tough Durable High tensile strength	F1 racing car bodies Road and mountain bikes Tennis racquets
Kevlar	Thermosetting	Easily moulded into shape Lightweight Tough Durable High tensile strength	Bulletproof/stabproof vests Crash helmets and motorcycle safety clothing
Extruded polystyrene foam (XPS) 'Styrofoam'	Thermoforming	Lightweight Easy to work Good thermal insulation	Thermal insulation in the construction industry Modelling
Epoxy resin (ER)	Thermosetting	Rigid Brittle Good thermal and electrical insulator Good chemical and wear resistance	Moulds for casting Adhesives Circuit boards

Finishing polymers

- Polymers are self-coloured, durable and waterproof and usually come with a high-gloss finish.
- There are ways of changing the finish if required; see Section 4, Topic 11.

see Section 4, Topic 11.

Exam tip

Be able to name several polymers, give their properties and suggest possible uses.

Now test yourself

TESTED

1. State two uses of Styrofoam and explain the properties that make it suitable for each purpose. [4]
2. Explain the difference between a thermoforming and a thermosetting polymer. [2]
3. Explain what is meant by the term 'super calendering'. [2]
4. Give one example of how a metal can be tested to find out if it is ferrous or non-ferrous. [2]
5. Name a suitable natural timber for use as a wooden spatula and explain your choice. [3]

5 Modern and smart materials

Quantum tunnelling composite (QTC)

- Quantum tunnelling composites are flexible polymers that contain conductive nickel particles that can be either a conductor of electricity or an insulator.
- The nickel particles make contact with each other and are compressed when force is applied, leading to an increase in conductivity. When the force is removed, the material returns to its original state and becomes an electrical insulator.

Polymorph

- For more on polymorph, see Section 1, Topic 4.

Thermochromic and photochromic pigments

- For details of these pigments, see Section 1, Topic 4.

Nitinol

- Nitinol is a shape memory alloy, made from titanium and nickel. Products made from nitinol can be shaped and deformed, but when heated the metal will return to its original state.
- Nitinol has many uses in medicine, for example the medical fastenings that are used in bone fractures, stents for heart surgery and dental implants such as braces.

6 Sources, origins, physical and working properties

Metals

Metals can be classified into three distinct categories:

- ferrous (containing iron)
- non-ferrous (does not contain iron)
- alloys (a metal made from two or more other metals).

Most metals are found in the Earth's crust and are embedded in rock known as ore.

- Ore can be opencast mined, underground mined or even dredged from rivers.
- Iron comes from an ore called **haematite**, which can be found in countries such as Brazil, Australia and South Africa.
- Ore must be smelted to release the metal from the rock.
- Iron is extracted from haematite by a process called **smelting**. This involves heating the ore to a very high temperature in a blast furnace.
- Aluminium is smelted from its ore (**bauxite**) in a reduction cell.

> **Haematite:** ore containing iron.
>
> **Smelting:** the process of extracting metal from ore.
>
> **Bauxite:** ore containing aluminium.

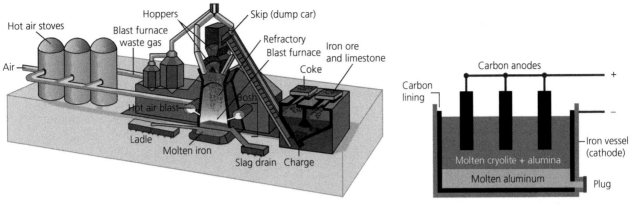

Figure 4.2 **The blast furnace**

Figure 4.3 **The reduction cell**

Ferrous metals

REVISED

All ferrous metals contain iron and when alloyed (mixed) with carbon they produce steel.

Table 4.10 **Common ferrous metals**

Ferrous metal	Composition	Properties	Common uses
Cast iron	Iron and 3.5% carbon	Hard surface but has a brittle soft core Strong compressive strength Poor resistance to corrosion 1200°C melting point Good electrical and thermal conductivity Cheap	Vices, car brake discs, cylinder blocks, manhole covers
Mild steel	Iron and 0.15–0.35% carbon	Good tensile strength, tough, malleable Poor resistance to corrosion 1500°C melting point Good electrical and thermal conductivity Cheap	Car bodies, nuts, bolts, and screws, RSJs and girders
Medium-carbon steel	Iron and 0.35–0.7% carbon	Good tensile strength Tougher and harder than mild steel Poor resistance to corrosion 1500°C melting point Good electrical and thermal conductivity	Gardening tools and springs
High-carbon steel	Iron and 0.70–1.4% carbon	Hard but also brittle Less tough, malleable or ductile than medium-carbon steel Poor resistance to corrosion 1500°C melting point Good electrical and thermal conductivity	Screwdrivers, chisels, taps and dies

Heat treatment of ferrous metals

The properties of ferrous metals can be altered by the use of heat.

- **Annealing** involves heating the metal to red heat and then allowing it to cool very slowly. This makes the metal as soft as possible.
- To **harden** ferrous metal, you must heat it to red heat and then cool it as quickly as possible in cold water.
- **Tempering** involves heating the steel to a known temperature and then allowing it to cool naturally. This removes the brittleness of hardened steel.
- The process of case hardening can harden the surface of a ferrous metal. The metal is heated to red heat and then placed in a high-carbon compound where it soaks up some of the carbon. This carbon-rich coating can then be hardened by heat treatment.

> **Annealing**: A method of heat-treating metal that makes it as soft as possible.
>
> **Hardening**: a method of heat-treating metal that makes it hard but brittle.
>
> **Tempering**: a method of heat-treating metal that reduces brittleness.

Non-ferrous metals

REVISED

Non-ferrous metals do not contain iron.

Table 4.11 **Common non-ferrous metals**

Non-ferrous metal	Composition	Properties	Common uses
Aluminium	Pure metal	Lightweight, soft, ductile and malleable A good conductor of heat and electricity Corrosion resistant 660°C melting point	Aircraft bodies, high-end car chassis, cans, cooking pans, bike frames
Copper	Pure metal	Extremely ductile and malleable An excellent conductor of heat and electricity Easily soldered and corrosion resistant 1084°C melting point	Plumbing fittings, hot water tanks, electrical wire
Silver	Pure metal	A soft, precious metal that is extremely resistant to corrosion An excellent conductor of heat and electricity 961°C melting point Expensive	Often used as jewellery

Heat treatment of non-ferrous metals

The properties of non-ferrous metals can also be altered by the use of heat. The main difference is that hardening and annealing non-ferrous metals happens at a much lower temperature.

Alloys

REVISED

An alloy is a metal that is produced by combining two or more elements to produce a new metal with enhanced properties.

Now test yourself answers at **www.hoddereducation.co.uk/myrevisionnotes**

Table 4.12 Common alloys

Alloy	Composition	Properties	Common uses
Stainless steel – ferrous alloy	Alloy of steel also including chromium (18%), nickel (8%) and magnesium (8%)	Hard and tough Excellent resistance to corrosion 1510°C melting point Good electrical and thermal conductivity	Sinks, cutlery, surgical equipment, homewares
High-speed steel – ferrous alloy	A medium-carbon alloy that also contains tungsten, chromium and vanadium	Very hard Resistant to friction Can only be sharpened by grinding 1540°C melting point Good electrical and thermal conductivity	Lathe cutting tools, drills, milling cutters
High-tensile steel – ferrous alloy	A low-carbon steel that also contains chromium and molybdenum	A steel with a very high yield strength 1540°C melting point	Used to reinforce concrete
Brass – non-ferrous alloy	Alloy of copper (65%) and zinc (35%)	Strong and ductile Casts well Corrosion resistant 930°C melting point Conductor of heat and electricity	Castings, forgings, taps, wood screws
Bronze – non-ferrous alloy	Copper 80–90%, tin, aluminium, phosphorous and/or nickel in varying amounts	Reddish-yellow in colour Harder than brass Corrosion resistant 1200–1600°C melting point	Castings, bearings and gears
Pewter – non-ferrous alloy	Tin 85–90%, antimony, copper	Malleable with a low melting point (170–230°C) Easy to work	Castings, beaten metalwork
Duralumin – non-ferrous alloy	Alloy of aluminium (90%), copper (4%), magnesium (1%), manganese (0.5–1%)	Strong, soft and malleable Excellent corrosion resistance Lightweight 660°C melting point Conductor of heat and electricity	Aircraft structure and fixings, suspension applications, fuel tanks

> **Exam tip**
>
> Make sure that you know the difference between a ferrous and a non-ferrous metal. Understand what is meant by the term 'alloy'. Be able to correctly name a number of metals and be aware of their properties and uses.

Natural and manufactured timber

Natural timber is categorised into two groups: **hardwoods** and **softwoods**.

Primary sources

It is vital that designers and manufacturers know where timber comes from and understand the processes that it has gone through before they begin to use it.

- Softwoods mainly come from cool northern parts of Europe, Canada and Russia.
- Hardwoods are grown in Central Europe, West Africa, Central and Southern America.
- When trees reach maturity they can then be cut down (felled) and **converted** into planks.
- Newly-felled timber contains a lot of moisture and is known as **green timber**. The process of removing moisture from newly-converted planks is known as **seasoning**.
- Timber is a natural product that has the benefit of being renewable.

Types of natural timber and their working properties

By understanding the properties of different types of timber you will be able to make an informed choice of which one to use when designing and making products. You can read about the working properties of different types of timber in Tables 4.4 and 4.5 (Section 4, Topic 2).

Available forms of natural timber

Natural timbers (hardwoods and softwoods) are generally supplied in planks, boards, strips and squares.

For more on the way in which timber is supplied, see Section 4, Topic 8.

Types of manufactured timber and their working properties

Manufactured boards are commercially produced sheets of timber that offer advantages over natural timber:

- They are available in much larger sheets than solid timber (2440 mm × 1220 mm).
- They have consistent properties throughout the board.
- They are more stable than natural timbers, meaning they are less likely to warp, shrink or twist.
- They can make use of lower-grade timber, so can have environmental and economic benefits.
- They can be faced with a veneer or a laminate to improve their aesthetic appearance.
- Due to their consistent quality, they are well suited to CNC machining and volume production.

> **Hardwoods:** timber that comes from deciduous trees and is generally harder than softwood.
>
> **Softwood:** timber that comes from coniferous trees and is generally less expensive than hardwoods.
>
> **Conversion:** the process of cutting up a log into planks.
>
> **Green timber:** timber that has just been felled and contains a lot of moisture.
>
> **Seasoning:** the process of removing moisture from newly-converted planks.

Manufactured boards fall into two categories:

● Laminated boards are produced by gluing together large sheets or veneers.
● Compressed boards are manufactured by gluing together particles, chips or flakes under pressure.

Table 4.13 **Manufactured timber**

Manufactured board	Description	Properties
Medium-density fibreboard (MDF)	Made from compressed fine wood fibres bonded together with resin	This board is relatively inexpensive and has a flat, smooth surface
Plywood	Made from wood veneers glued together with alternating grain	Very strong, with a flat, smooth surface
Chipboard	Made from wood chips bonded together with resin	Inexpensive construction material with limited strength
Hardboard	Made from compressed fine wood fibres bonded together with resin. Has one smooth side and one textured side	Very inexpensive material used for drawer bases and backs of wardrobes

Exam tip

Learn the names and the properties of several natural timbers and manufactured boards. Make sure that you can suggest possible uses for them and be ready to identify them from a photograph.

Thermosetting and thermoforming polymers

REVISED

For details of the differences between thermosetting and thermoforming polymers and the physical and mechanical properties of different types of polymers, see Section 4, Topic 4.

Synthetic polymers

● Most polymers are made from crude oil and are known as synthetic polymers.
● Crude oil is found throughout the world, with the biggest deposits being in the Middle East and in Central and South America.
● It is extracted from the ground by drilling and pumping it to the surface. It is then transported to an oil refinery to be processed.

Natural polymers (biopolymers)

● Biopolymers are made from plant-based materials such as sugar beet and corn starch.
● Being plant based means that biopolymers are renewable.

Additives

There are a number of additives that can be blended with polymers to enhance certain properties.

● Polymerisers are added to polymers to enhance their flexibility.
● Pigments change the colour of a polymer.
● Fillers are added to polymers to increase their bulk and reduce their cost.
● Flame retardants can be added to polymers to prevent or slow down the rate at which they burn.

Exam tip

Make sure that you know the difference between a thermoforming polymer, a thermosetting polymer and a natural polymer (biopolymer).

Papers and boards

Sources

- The first paper was made from fibres of tree bark mixed in water, known as pulp. The pulp was drained, spread out and pressed down into a thin layer before being dried out in the sun.
- It was later discovered that lignin, the natural glue that holds the wood's fibres together, could be broken down more easily if plants with long cellulose fibres were used. This meant the fibres could be made into a finer pulp which made better quality paper.
- It takes around 12 average-sized trees to make 1 tonne of newspaper and around 24 trees to make 1 tonne of copier paper.

> **Exam tip**
>
> Papers and cardboard are made from trees and therefore products made from them have a **life cycle** similar to timber products.

> **Life cycle**: the stages a product goes through from beginning (extraction of raw materials) to end (disposal).

Recycled paper

- Recycled paper reduces the number of trees needed and lessens the environmental impact.
- Paper cannot be recycled indefinitely as after around five or six times, the fibres become too short and weak to pulp adequately.
- To maintain the quality and strength of recycled paper, a mixture of recycled paper and new virgin wood chippings is used to make the pulp (55–80% recycled paper and 20–45% virgin wood chippings).

Table 4.14 **Physical and working properties of paper**

Common name	Weight (gsm)	Properties/working characteristics	Uses
Layout paper	50	Bright white, smooth, lightweight (thin) so slightly transparent and inexpensive	Sketching and developing design ideas, tracing parts of designs
Copier paper	80	Bright white, smooth, medium weight, widely available	Printing and photocopying
Cartridge paper	80–140	Textured surface with creamy colour	Drawing with pencil, crayons, pastels, water colour paints, inks and gouache
Bleedproof paper	80–140	Bright white, smooth surface, stops marker 'bleed'	Drawing with marker pens
Sugar paper	100	Available in wide range of colours, inexpensive, rough surface	Mounting and display work

Board

Card

- One method of making cardboard is to sandwich and paste together multiple layers of paper.
- Another method is to press together the layers of wet pulp into a thicker layer.

Corrugated card

- Corrugated cardboard has three layers made by passing paper through a corrugation machine.
- The centre layer is steamed to soften the fibres, then crimped to give it a wavy shape.
- The two outer layers of paper are then glued on each side of the wavy centre layer.
- After corrugation, it is cut into large pieces or 'blanks' which then go to other machines for printing, cutting and gluing together.
- Double-walled corrugated card has an additional wavy and flat layer to make it more rigid and give extra protection.

Figure 4.4 **Corrugated card**

Table 4.15 **Physical and working properties of card**

Common name	Thickness (microns)	Properties/working characteristics	Uses
Card	180–300	Available in a wide range of colours, sizes and finishes, easy to fold, cut and print onto	Greetings cards, paperback book covers, as well as simple modelling
Cardboard	300 upwards	Available in a wide range of sizes and finishes, easy to fold, cut and print onto	General retail packaging such as food, toys, design modelling
Corrugated cardboard	3000 upwards	Lightweight yet strong, difficult to fold, good heat insulator	Pizza boxes, shoe boxes, larger product packaging, e.g. electrical goods
Mounting board	1400	Smooth, rigid, good fade resistance	Borders and mounts for picture frames

Laminated layers

REVISED

Foam board and Styrofoam

- Foam board and Styrofoam are slightly different types of polystyrene foam.
- Foam board uses this foam sandwiched between two outer layers of paper or thin card.
- It is commonly used for architectural modelling, prototyping small objects and large hanging signs in supermarkets.
- Styrofoam is a trade name for extruded polystyrene foam insulation (XPS).
- XPS is made by melting plastic resin and other ingredients into liquid form and extruding through a die.
- The extruded liquid then expands as it cools, producing a closed-cell rigid insulation.
- Styrofoam is used as a modelling material to create moulds for vacuum forming and as insulation.

Corriflute

- Corriflute is a trade name for corrugated polypropylene plastic.
- It is manufactured in one piece by extruding molten polypropylene through a former that moulds the polymer into the shape required.

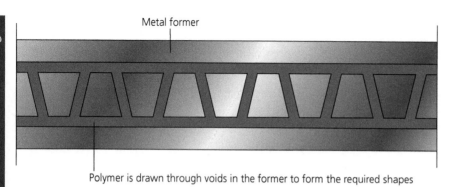

Metal former

Polymer is drawn through voids in the former to form the required shapes

Figure 4.5 **Section through Corriflute**

Foamex

- Foamex is a tradename for polyvinyl chloride foam.
- PVC foam is made by mixing two chemicals (diisocyanates and polyols) which react with each other.
- Pigments for colour and other additives are added.
- The mixture is then poured onto a moving conveyor belt where it 'foams', forming one long continuous block of foam.
- It is lightweight and rigid, with good insulation properties, and can be printed onto easily.
- It is available in a range of colours and is easy to cut and join to other materials.
- It is resistant to water and many chemicals so is often used for outdoor displays and signs.

> **Exam tip**
>
> Corriflute, PVC foam and Styrofoam are made from polymers and therefore have the same basic characteristics of plastics (e.g. waterproof, deform with heat).

Table 4.16 **Properties of types of board**

Common name	Properties/working characteristics	Uses
Foam board	Smooth, rigid, very lightweight and easy to cut	Point-of-sale displays, ceiling-hung signs in supermarkets, architectural modelling
Styrofoam	Light blue colour, easy to cut, sand and shape, water resistant, good heat and sound insulator	3D moulds for vacuum forming and glass reinforced plastic (GRP), wall insulation in caravans, boats
Corriflute	Wide range of colours, waterproof, easy to cut, rigid, lightweight	Outdoor signs, packaging and modelling
Foamex	Lightweight, good insulation properties, easy to print on, water resistant	Outside displays and signs

Laminating

REVISED

Laminating is usually done to finished documents to:

- improve their strength and resistance to bending, creasing or ripping
- waterproof the document, allowing it to be wiped clean and prevent it smudging or going soggy
- improve the appearance, making the document shiny
- increase the lifespan of the printed document.

Laminating involves applying a film of clear plastic between 1.2 mm and 1.8 mm thick to either one or both sides of paper or thin card. There are three methods of laminating a document:

Pouch lamination

- Pouch lamination uses thin, clear plastic pouches coated on the inside with a thin layer of heat-activated glue.
- The document is placed inside the film pouch then through a laminating machine.
- The machine heats the pouch, activating the glue which seals the pouch together as it is pressed through the rollers, encasing the document inside it.
- Pouch laminators can do only one document at a time so are ideal for small-volume items.

Thermal or hot lamination

- Where large volumes of laminated documents are required, commercial lamination methods are required. Thermal lamination uses rolls of thin, heat-sensitive plastic film.

Cold laminating

- Cold laminating is done when only one side of the paper or card is to be coated.
- Cold laminating does not use heat, making it ideal for documents such as photographs.

7 Selection of materials and components

Metals

REVISED

- Metals are selected for both their functional and aesthetic properties.
- They are more difficult to work than most materials due to their inherent hardness but can be manufactured to greater accuracy.

Table 4.17 **Functional and aesthetic properties of ferrous and non-ferrous metals**

Metal	Aesthetics	Functionality
Aluminium (Duralumin)	Easily cast into unique shapes, can be polished to a mirror-like finish or be coloured with a vivid finish known as anodising	Excellent strength-to-weight ratio, easy to cut, weld and join by various methods, good resistance to corrosion
Copper	Easy to shape by beating, has a reddish-brown finish that can be highly polished. Has the unique feature of going green when left outdoors and unprotected	Easily worked, good conductor of heat and electricity, malleable, ductile and easily joined by soldering
Brass	Easily cast into unique shapes, has a yellowy-brown colour that can be highly polished	A harder, more durable material than copper, a good conductor of heat and electricity
Pewter	Very easily cast into shape due to its low melting point	Soft metal with a relatively low strength
Mild steel	Can easily be worked into shape, rusts if left unprotected, can receive a wide variety of finishes/platings/coatings	Tough, durable, strong and malleable, relatively easy to work

Environmental factors

- Metals come from ore, which is a non-renewable resource.
- The processing of ore into pure metal uses a lot of energy. Most of this energy comes from non-renewable fossil fuels.
- Some metals will pollute the ground if they go to landfill.
- Most metals are relatively easily to repair, have a good life span and can be recycled.

Availability

- Metal is readily available in a variety of stock forms (see Section 2, Topic 8).
- Metal comes in stock sizes of length, width, thickness (gauge), diameter and weight.

True cost

- The cost of metals can vary quite considerably. Common metals such as steel are relatively inexpensive. Semi-precious metals such as copper, tin and lead are more expensive, while precious metals such as gold and silver are very expensive.
- Metals are suitable for batch, mass and continuous methods of manufacture, which reduces the unit cost of everyday products such as aluminium drinks cans.
- Bespoke items of jewellery, made by hand using precious metals, are very expensive.

Social and cultural factors

- Constructional metals such as steel are very accessible and affordable and therefore they are extensively used by all.
- Precious metals such as gold and silver are affordable only by the wealthy members of society.

Ethical factors

- Metals are a finite resource and must be recycled at the end of their life or they will not be available for future generations.
- The processing of metal ore pollutes the planet.
- The ELVD (End of Life Vehicles Directive) is a European directive aimed at ensuring that vehicles are correctly dealt with at the end of their usable life.

Timber

REVISED

Functionality

- A garden bench needs to be strong, durable and weatherproof.
- A child's toy needs to be tough and free from splinters.
- Low-cost furniture needs to be flat, stable and easy to assemble at home.
- Different timbers have different properties:
  - Pine is a relatively inexpensive natural timber that has an attractive grain and is available in long straight lengths, making it ideal for larger prototypes.

- ○ Beech is a very tough, durable timber that does not splinter, making it a suitable material for use in children's toys.
- ○ Mahogany is a strong timber with a deep red colouring that can be polished to a very high-gloss finish. It is often used to manufacture smaller projects such as jewellery boxes.

Aesthetics

- There is a range of wood colours from sycamore (very pale) to ebony (black).
- Timber can be stained, painted and have a gloss, satin or matt finish applied to it.
- Timber is very tactile; it has a natural grain that can be sanded smooth or left quite rough.
- Timber is 'warm' to the touch.

Environmental factors

- Timber is environmentally friendly as it can be renewed by growing more trees. Products made from natural timbers are relatively easy to repair, the timber can be recycled and will not harm the environment if it goes to landfill.
- Timber is biodegradable and will not harm the environment when it is disposed of.
- It is relatively easy to repair and most timbers have a good lifespan.
- Manufactured boards are less environmentally friendly as they have undergone additional manufacturing processes and many contain adhesives that make recycling difficult.

Availability

- Softwoods are relatively quick growing and there is an abundance of renewable timber.
- Hardwoods take longer to grow.
- Most timber comes in stock sizes, which makes designing and making easier to plan.
- Manufactured boards come in large stock sizes and in a variety of thicknesses.

True cost

- Softwoods are a relatively low-cost natural timber. This is because we have an abundant supply of softwood, it is relatively quick growing and easily converted into usable planks.
- Many manufactured timbers are also low-cost. This is because they can be made from low-grade or recycled timber and are produced in high quantities.
- Hardwoods tend to be more expensive since they are slower growing and require a more involved process to convert them into a usable material.
- Some exotic hardwoods, such as burr walnut, can be very expensive. These are specialist timbers that are difficult to source and require expert attention to turn them into high-quality usable timber.
- Labour-intensive manufacturing costs of bespoke handmade furniture can increase the price of wooden products.

- Manufactured boards are suitable for mass production by CNC machinery that significantly reduces the cost of making.
- Manufactured timber is available in large sheets (2440 mm × 1220 mm) and in a variety of thicknesses, making it ideal for use in cost-effective, self-assembly furniture.
- Bulk buying of timber can reduce the cost of the material.

Social factors

- Timber is an accessible, affordable material.
- It is used in all countries as a building and construction material, providing inclusive housing and furniture for everyone.

Cultural factors

- Different cultures have different needs and tastes regarding timber-based products.
- The Japanese make extensive use of bamboo as a timber-based construction material.
- In Northern Europe, log cabins are made from spruce.

Ethical factors

- Timber creates few ethical issues as it is a natural, renewable material.
- If timber forests are not managed, this can lead to deforestation.
- Deforestation can cause the loss of habitats for wildlife and can be a contributor to global warming.

Biodiversity

- Forests are a very biodiverse environment.
- They provide a habitat for many types of plants and wildlife.
- Animals, birds, insects, grasses and flowers live within the forest and rely on them for their existence.

Paper and card

REVISED

Aesthetic and functional properties

- The main influence on selection of material will be the properties and characteristics of the material. The following properties need to be considered:
  - Surface finishes: these can affect strength, rigidity and gloss (shiny) or matt finish.
  - Absorbency: the lower the absorbency, the higher the quality of the image when printed onto.
  - Colour: many products require a specific colour paper to be used.
  - Texture: the rougher the texture, the better the material is to draw or paint on.
  - Flexibility/rigidity: some products need to bend whereas others require a rigid, stiff material.
  - Water resistance: certain outdoor applications require waterproof boards such as Corriflute.

- Modern printing methods can print onto a wide range of materials and different-shaped products. Traditional printing methods cannot print onto materials such as corrugated card, foam board or Corriflute.
- Low manufacturing costs maximise profit, making the product as cheap to buy as possible.

Availability

- Products may require sizes, colours or textures that are non-standard and have to be made to order.
- These take longer to produce, are made in limited quantities and are more expensive.

Environmental influences

- Due to the impact of paper making on **deforestation**, many paper manufacturers use only trees from **managed forests** such as those run by the **Forest Stewardship Council (FSC)**.

Water and energy consumption

REVISED

- Paper production can lower the water table due to the amount of water required and increase water temperature and sedimentation, which affects local wildlife.
- Paper-making machines also use large amounts of electricity.
- Many paper mills now recycle up to 90 per cent of the water used.
- Chemical and solvents: unbleached paper can be used for products that do not need to be white in colour by reducing toxic solvents and chlorine.
- Air pollution: pulp and paper mills produce carbon dioxide and other pollutants that damage the ozone layer, cause acid rain and contribute to global warming.
- Solid waste: waste paper fibres form a sludge which is either disposed of into landfill or can be dried out and burned as a fuel.

> **Deforestation**: the removal of trees from an area of land which are not replanted.
>
> **Managed forest**: a forest where new trees are planted whenever one is cut down.
>
> **Forest Stewardship Council (FSC)**: organisation that promotes environmentally appropriate, socially beneficial and economically viable management of the world's forests.

Social, cultural and ethical issues

REVISED

- When designing graphic products, social, cultural and ethical issues must be considered.
- Imagery, symbols and even certain colours can mean different things in other **cultures**.
- Designers must take care over how they portray individuals or groups of people on products and when selecting and using photographs or images of people on their products. Careful thought must go into selecting images and consideration of how an image may represent a minority group.
- Using a person of a particular race can sometimes portray them in a negative light or stereotypical manner which can be extremely offensive. Designers must consider different social attitudes and how these differ across the world. What is socially acceptable in the western world may not be in other countries.
- In other cultures there are different beliefs and attitudes regarding revealing clothing and keeping certain parts of the body covered at all times.

> **Culture**: the ideas, customs and social behaviours of a particular people or society.

Responsibilities of designers

- Due to globalisation, many products are produced in developing countries where labour rates and materials are much cheaper.
- In the western world, health and safety laws protect the safety, rights and welfare of all workers. Developing countries have less stringent safety laws and poor people can be exploited.
- **Exploitation** is when workers are forced to work:
  - in unsafe, unhealthy or dangerous conditions
  - for extremely long hours without sufficient breaks
  - without correct protective equipment
  - for low pay rates that do not reflect a fair wage.
- Designers have a responsibility to 'refuse' to design products for companies that exploit their workers in this way.
- Organisations such as Fairtrade ensure that workers are not exploited in this way.
- Consumers can support the rights of workers by refusing to purchase goods not approved by these organisations.
- Designers have a responsibility to design products that are as eco-friendly as possible.
- Using recycled paper and card whenever possible and designing products that can be easily recycled is one way of doing this, for example avoiding the use of coatings that make recycling paper and card products difficult or the use of vegetable-based inks instead of solvents.

> **Exploitation:** treating someone unfairly in order to benefit from their work.

Estimating the true costs of a prototype or product

- The price of producing a prototype or final product will vary depending on many factors, such as:
  - the quantity required
  - the size and complexity of the product
  - the materials and processes used to make it
  - the production time
  - the price and availability of any manufactured components.
- Designers should carefully consider and calculate the amount of material needed.
- The costs of paper and boards reduce the more you purchase.
- The cost per mm^2 of all sheet materials decreases as the sheet size increases.
- It is more economical to buy a large-size sheet and cut it down into smaller sheets than buy it ready cut.
- Large sheets are more difficult to transport so the designer must also factor in transport and delivery costs.
- Buying materials from countries on the other side of the world can often be cheaper than buying in your own country.
- The effects of globalisation and local economies must be considered when deciding where to purchase products.

8 Stock forms, types and sizes

Natural timber   REVISED

- Timbers are readily available as planks, boards, strips, dowels, mouldings
  and square sections.

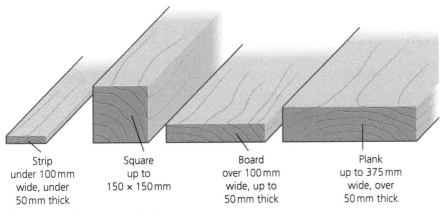

Strip	Square	Board	Plank
under 100mm	up to	over 100mm	up to 375mm
wide, under	150 × 150mm	wide, up to	wide, over
50mm thick		50mm thick	50mm thick

Figure 4.6 **Timber stock forms**

- Natural timber that has come straight from the sawmill is known as
  rough sawn. It is generally used for construction work where appearance
  is not important.
- Timber that has had just its sides planed is known as **planed both sides
  (PBS)**.
- Timber that has had all sides planed is known as **planed square edge
  (PSE)** or **planed all round (PAR)**. This can be used for a wide variety of
  applications, including interior joinery.

> PBS: planed both sides.
>
> PSE: planed square edge.
>
> PAR: planed all round.

Manufactured boards

- Manufactured boards such MDF, plywood, hardboard and
  chipboard come in large sheets: 2400 mm × 1200 mm.
- The thickness of manufactured boards can range from
  1 mm modelling plywood to 40 mm MDF, suitable for the
  core of a kitchen worktop.

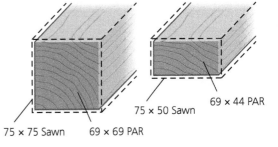

75 × 75 Sawn   69 × 69 PAR
75 × 50 Sawn   69 × 44 PAR

Figure 4.7 **Typical timber planed sizes**

Mouldings and dowelling

Timber can also be supplied in a wide range of decorative mouldings.
These are very useful for products such as picture frames. Dowelling is
timber that is cylindrical and is useful for making wooden rails.

Veneers

Veneers are thin sheets of wood that can be used for constructional or decorative use.

- Construction veneers are typically used in the manufacture of plywood.
- Decorative veneers can be applied to manufactured boards to enhance their appearance.

Calculating the cost involved in designing timber-based products

- In addition to the cost of buying the timber you should consider the cost of purchasing fixtures and fittings such as hinges, screws and handles.
- Different types of finish can also affect the final cost of a product. A simple wax finish is relatively cheap to buy and quick to apply, whereas a high-quality finish such as French polish involves an intricate process that requires a high degree of skill and is much more time consuming, making it significantly more expensive.

Polymers

REVISED

- Polymers are available in a wide range of forms, colours and thicknesses.
- A big advantage of using polymers is that they can be bought self-coloured with an immaculate high-gloss finish.

Table 4.18 **Stock forms of polymers**

Stock form	Description
Sheet	The most commonly used sheet polymer in the school workshop is acrylic (PMMA). The sheet comes in regular sizes measuring 1200 mm × 600 mm × 3 mm
	Other polymer sheet materials include high impact polystyrene (HIPS). This is available in a size that fits most vacuum-forming machines, 475 mm × 274 mm × 1 mm
Extrusions	Typical extrusions include rod and tube. Specific extrusions include products such as curtain rails
Granules/pellets	Mainly used in industrial processes such as injection moulding and extrusion
Powders	Used in the fluidising bath as part of the plastic dipping process
	In industry they would be used in powder coating and also with thermosetting polymers in compression moulding
Foams (closed cell)	Available in large sheets (1200 mm × 600 mm) and extensively used in school for model making
Foams (open cell)	Usually obtained in a can and sprayed into a cavity. Have good thermal, buoyancy and sound-insulating properties
Film	A very thin sheet that is usually sold in rolls
Filament	Polylactic acid (PLA) is sold in rolls and is used in many rapid prototyping machines
Liquid	A number of polymers are available in a liquid form. Epoxy resins are sold in liquid form and become solid only when mixed with a chemical catalyst

Calculating the cost of polymer-based products

Calculating the cost of a polymer-based product is influenced by the type of polymer and the method of production.

- The cost of products made from PMMA that is to be laser cut will be calculated from the surface area of PMMA needed plus the power and time consumed by the laser cutter.
- The cost of a prototype produced in PLA by a 3D printer will be calculated by the length of filament needed plus the power and time consumed by the 3D printer.
- The cost of a product produced on an industrial injection-moulding machine from polypropylene (PP) will be calculated from the volume of PP needed plus the power and time consumed by the injection-moulding machine.

Figure 4.8 Polymer extrusions

Paper and board sizes

REVISED

Figure 4.9 Polymer granules

- Paper and board sizes range from A10 through to 4A0.
- The most common sizes used by designers are between A6 and A0.
- Foam board is available in standard sizes from A4 to A0 and thicknesses of 3 mm, 5 mm or 10 mm. Many suppliers stock standard imperial sizes up to 8 ft × 4 ft (2440 mm × 1220 mm).
- Corriflute is available in a range of standard sizes. It comes in thicknesses of 2–10 mm and in a range of colours.
- PVC foam is available in standard paper sizes and thicknesses of 1–6 mm, 8 mm, 10 mm, 13 mm, 15 mm, 19 mm and 25 mm. A range of standard and special designer colours with gloss or matt finishes is available.
- Styrofoam is available in sheet form or in blocks. Sheet thicknesses range from 5 mm to 165 mm through increments of 5 mm or 10 mm. Thicknesses above 165 mm are considered a block rather than a sheet.

Costs

- Paper products can be costed by multiplying the cost per sheet by the number of sheets required.
- Paper is much cheaper per sheet when bought in **reams**.
- Card products can be costed in the same way.
- The size of card required can be calculated by finding the optimum sheet size for the number of items required (see 'tessellation' in Section 4, Topic 9).
- Corriflute, foam board and Styrofoam costs can be calculated by working out the combined surface area of all the pieces required and the minimum standard sheet size they will fit onto.

Ream: pack of 500 sheets.

9 Manufacturing to different scales of production

Scales of production

One-off production

- **One-off production** is used to manufacture single bespoke products.
- Products are manufactured by highly skilled workers.
- Products are generally very expensive.
- The production method is labour intensive and time consuming.
- More expensive materials such as high-quality, exotic timbers and precious metals can be used.

> **One-off production:** a process used when making a prototype product.
>
> **Batch production:** making a small number of the same or a similar product.

Batch production

- With **batch production** a limited number of identical products can be made to a high consistency.
- Materials can be purchased in bulk, thereby reducing the cost.
- Machinery can be set up to manufacture in quantity, saving time.
- Less skilled labour is required.
- The unique appeal of a 'one-off' product is lost.

The use of jigs

- A jig is a device that is specially made to perform a specific part of the manufacturing process.
- Jigs are useful when the process has to be carried out multiple times. They can be used when cutting, drilling, sawing and gluing.
- They have a number of very important advantages:
  - They speed up the manufacturing process.
  - They reduce the risk of human error.
  - They reduce the unit cost of a part.
  - They make the process safer to carry out.
  - They increase the accuracy of the process.
  - They increase the consistency of the process.
  - They reduce wastage.
- There are some disadvantages of using jigs:
  - They are only cost effective when large numbers of similar parts are required.
  - They increase the initial cost of the part.
  - They require a high level of skill to produce.

Mass production

- **Mass production** allows metal products to be produced in large quantities.
- Bulk buying of materials significantly reduces cost.
- Specialist machinery and the increased use of CAM are essential features of high-volume production. They increase consistency, accuracy and the speed of production.
- An unskilled/non specialist workforce can be used.

> **Mass production:** large quantities of identical products are produced.

Now test yourself answers at **www.hoddereducation.co.uk/myrevisionnotes**

High-volume/continuous flow production

- Products that are in very high demand are made continuously for 24 hours per day, seven days per week.
- Highly specialised equipment and extensive use of CAM are part of the process to manufacture products.
- The process can be fully automated, deskilling the workforce, who become involved in servicing and maintenance engineering.
- This requires a large initial investment and is suitable only where there is a high demand for the products.
- Water bottles and tin cans are examples of products made by continuous production.

Issues with high-volume production

High-volume production creates a number of issues.

- Workers become deskilled and there is less employment as the machines take over manufacture.
- Products become very similar and lose their uniqueness.
- More energy is need to power factories, creating greater pollution for our planet.

Computer-aided manufacture

- CAM is used when products are manufactured in high volumes.
- This requires a large initial investment in new factories and state-of-the-art machines.
- It is cost effective only when there is a very high demand for identical products.

Processes in manufacturing polymer products

There are a number of processes that are used to manufacture polymer-based products in high volumes. These are:

- injection moulding
- vacuum forming
- press forming
- compression moulding.

All these processes use the same principles of manufacturing.

- The polymer is heated and becomes soft and pliable.
- The polymer is then blown, sucked, drawn or pressed into a die or mould.
- The polymer takes the form of the die or mould.
- The polymer is cooled.
- The product is removed from the die or mould and then trimmed and finished.

On-press and the finishing processes used by commercial printers

- Once a design has been drawn on paper or produced on a software program, it must go through pre-press stages before it can be printed.
- The first stage checks that the fonts and formatting are correct.
- Then the resolution (sharpness) and colours of images are checked to ensure the four separately printed CMYK colours produce the required colour when printed.
- The layout of the page is checked to ensure it fits on the page and in the correct position:
  - Registration marks printed on the edge of the page are used to line up the different letterpress plates for multi-colour printing jobs.
- The imposition checks the arrangement of pages on the printer's sheet is optimised to allow faster printing, simplify binding and reduce paper waste.
- Once all the checks have been made, a 'proof' of the document is created and sent to the client for final approval.

Techniques to produce printed products

REVISED

Digital printing

- Digital printing is the most cost-effective batch production method. Digital printers are inexpensive to buy, readily available and many people have them for household use.
- The cost per sheet of printing using a digital printer is high compared with other types of printing but they are the best option for small print runs.

Screen printing

- Screen printing is another method of printing suitable for batch production.
- Screen printing is used for creating repeating patterns or designs. The lightest colour is printed first, then the screen is washed and masked up for the next colour. The process is repeated until all the colours have been printed.

Offset lithography

- This is one of the most common forms of commercial printing.
- Four ink colours are used: cyan, magenta, yellow and black (CMYK).
- The colours are overlaid to create others, for example cyan on top of yellow creates green.

Flexography

- This is another type of mass-production printing process.
- The process is quicker and cheaper than lithography, although the quality is not as good.
- It is used for printing on packaging where the quality of print is not as important.

Stencils and templates

- Stencils and templates can be used for batch production of paper and board products.
- The template is laid onto the material and drawn around, then moved and drawn around again and again until the required number is reached.
- A template ensures that every piece is exactly the same and saves the time of drawing out each part individually.
- **Tessellation** reduces waste by arranging the pieces in such a way that the material being cut out is used to its maximum capacity.

10 Specialist techniques and processes

> **Tessellation:** an arrangement of shapes closely fitted together in a repeated pattern without gaps or overlapping.

Wastage and addition

REVISED

- **Wastage** is the process of shaping material by cutting, sawing and shaping materials to form a desired shape.
- Addition is the adding of material by joining components together in some way.

> **Wastage:** the process of shaping material by cutting away waste material.

Timber

REVISED

Marking out

- Before marking out onto a piece of timber you should ensure that you have one face and one edge planed smooth. These are known as the 'face side' and the 'face edge'.
- A pencil and a ruler are used to mark out wood. A marking knife will give a more accurate cut line.
- A try square will give you an accurate 90° line to an edge. It will also allow you to check to see if an angle is at 90°.
- A mitre square produces an accurate 45° angle and a sliding bevel can be set to any angle.
- A marking gauge will produce a line parallel to an edge while a mortice gauge will produce a double parallel line.
- A template can be used as a method of marking out irregular shapes on wood. You can also draw around it to use it as an aid to quantity production.

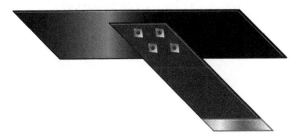

Figure 4.10 **A mitre square**

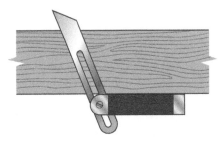

Figure 4.11 **A sliding bevel**

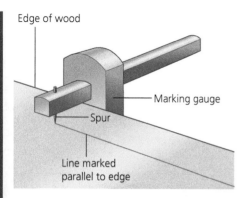

Figure 4.12 **A marking gauge**

Figure 4.13 **A mortice gauge**

Sawing wood

- A tenon saw will cut straight lines in wood. Make sure the wood is firmly held in a vice or with a G-clamp.
- A hand saw is used for cutting large pieces of wood and a coping saw will cut curves.
- The coping saw can be difficult to use accurately and therefore you should always cut slightly away from your line.
- Bandsaws and scroll saws are mechanical saws that will speed up the process and can improve accuracy.

Shaping wood

- Wood can be easily shaped using a surform or wood rasp.
- A disc sander, belt sander and linisher are machines that use coarse glass paper to shape and smooth wood.
- Planes can be used to smooth and shape wood.
- A variety of drills is used for making holes in wood. A hole saw can produce large holes; a Forstner bit can produce flat-bottom holes of varying sizes.
- Chisels are used to shape wood but are also very useful when cutting joints. Care should be taken when chiselling as chisels are very sharp. Always have your hands behind the cutting edge and securely clamp your work.

Drilling wood

- Drills will produce a circular hole in timber.
- Drilling machines can be a permanent feature in a workshop or may be hand-held.
- Drill bits come in a wide variety of sizes.

Figure 4.14 **Sawing with a hand saw**

Figure 4.15 **Sawing with a tenon saw**

Figure 4.16 **Using a chisel to cut a housing joint**

Metals

Marking out

- A scriber acts as a pencil when marking out on metal. On bright metal it is useful to use layout fluid to make it easier to see the lines.
- An engineer's square produces a 90° line to an edge and a centre punch will mark the centre of a hole before drilling.

- Outside and inside callipers measure the outside and inside of bar and tubing.
- Digital micrometres and vernier callipers can measure to 1/100th of a millimetre.

Sawing

- The hacksaw is the most common saw used for cutting straight lines in metal. A junior hacksaw is a small version used for cutting smaller metal parts.
- Bandsaws, jigsaws and even a coping saw can all be fitted with metal cutting blades.

Shaping

- Metal can be shaped using a range of file types. Files are graded depending on their roughness and these are known as 'rough cut', 'second cut' and 'smooth cut'.
- **Cross filing** is an effective method of shaping metal. You should use the full length of the file and remember that the file only cuts on the forward stock.
- **Draw filing** smooths the edges of metal. Place the file across the edge of the metal while holding the blade rather than the handle. Then move the file forwards and backwards to achieve a flat, smooth edge.

Cross filing: a method of shaping metal using files.

Draw filing: a method of smoothing the edges of metal.

Figure 4.17 **Cross filing**

Figure 4.18 **Draw filing**

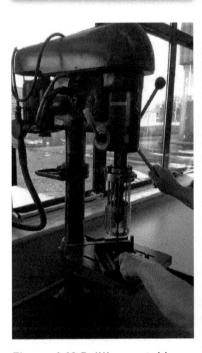

Figure 4.19 **Drilling metal in a machine vice**

Drilling metal

- Metal must be centre punched before drilling to ensure that the drill bit does not slip.
- The speed of the drilling machine should be set to the size of the drill and the hardness of the metal. Soft metals and small drills require a fast speed while hard metal and large drills need a slower speed.
- Drilling metal is normally done using a pillar drill, with the metal firmly held in a machine vice.
- Drill bits will produce a regular circular hole in metal.
- A countersink bit will produce a countersunk hole.
- A pilot hole is a small hole produced to guide a large hole.
- A tapping hole is produced before an internal screw thread is formed.
- A clearance hole is produced to allow a bolt to slip through.

Figure 4.20 **A countersink drill bit**

Countersunk hole   Counterbored hole

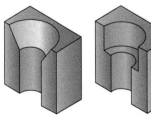

Figure 4.21 **A countersunk hole and a counterbored hole**

Polymers

Marking out

- A spirit-based pen will draw onto a polymer surface without scratching it.
- Paper templates allow a design to be transferred to the surface of the polymer and will give extra protection to the shiny surface.
- Keep the protective sheet on your polymer for as long as possible: it will help keep the surface clean and free from scratches.
- A laser cutter can be used to etch marking-out lines onto the surface of most polymers.

Holding polymers

- When working with polymer-based materials, you can hold them in a vice or with a G-clamp. Care should be taken to protect the surface.

Sawing

- Most metal and woodworking saws will cut polymers. The coping saw will allow you to cut around curves in acrylic.
- A scroll saw is a mechanised coping saw and will speed up the process.
- A craft knife will cut thin polymer sheets such as the HIPS sheets used for vacuum forming.

Shaping and forming polymers

- Polymers can be filed and sanded into shape by traditional methods using metalworking files and abrasive papers such as 'wet or dry' paper.
- Thermoforming polymers can be formed using a strip heater to produce a bend along a straight line.
- An oven can be used to soften a thermoforming polymer before it is shaped in a mould.
- Most cutting and shaping tools used in the manufacture of metal products can be used when making polymer-based components.
- A disc sander will speed up the process of shaping polymers.

Drilling polymers

- Polymers can be drilled using regular metalworking drill bits.
- Using a pillar drill will improve the accuracy of drilling but a cordless drill has the advantage of being able to drill in remote locations.
- Drilling a pilot hole will improve the accuracy of drilling and prevents the build-up of heat that would melt the polymer.
- Drilling a clearance hole allows the threaded part of the bolt to slip through the component before fastening with a nut.
- You may also use a tapping hole, countersunk hole or a counterbored hole – see 'Drilling metal' in Section 4, Topic 10 for details.

Figure 4.22 **A strip heater**

Figure 4.23 **A cordless drill**

Paper and boards

Marking out

- Pencil or pen can be used to mark out on paper, card and foam board.
- Styrofoam, Corriflute and PVC foam can be marked with a thin permanent marker or chinagraph pencil.

Cutting

- Scissors and craft knives are used for cutting paper and card.
- Foam board, PVC foam and Corriflute are too thick to be cut with scissors so must be cut using a craft knife.
- Styrofoam thicker than 10 mm can be cut using a serrated-edged knife, bandsaw, hacksaw blade or hot wire cutter.
- Final shaping can be done by sanding and smoothing with files and abrasive paper.
- Laser cutters can be used to cut any 2D shape out of card, PVC foam, foam board or Corriflute.
- Die cutters are used to crease, perforate and cut card for mass production.
- The 'die' is a set of sharp metal blades shaped to the outline of the net and fixed to a backboard.
- The card or paper being cut is placed on a flat surface and the die is pressed onto the material which cuts it to the required shape.

Deforming and reforming

Timber: joining wood

There is a wide range of methods of joining wood and these are categorised into carcase, stool and frame joints.

- Carcase joints are used to make box-type constructions. Joints such as the butt joint are quite simple to produce but are relatively weak. The dovetail joint is difficult to produce but provides very good strength.

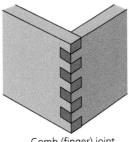

Comb (finger) joint

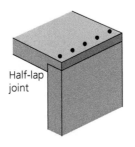

Dovetail nailing

Butt joint

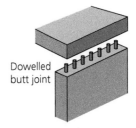

Dowelled butt joint

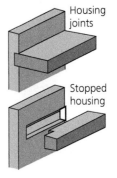

Housing joints

Stopped housing

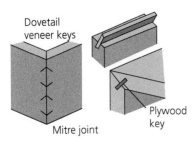

Dovetail veneer keys

Mitre joint

Plywood key

Half-lap joint

Figure 4.24 **Carcase or box joints**

- Stools, tables and chairs can be constructed using stool joints.

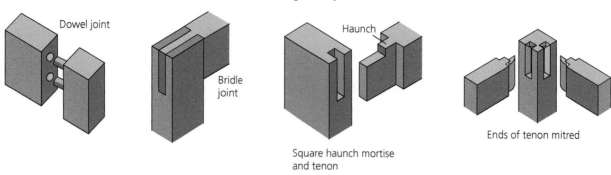

Figure 4.25 Stool joints

- Wooden windows and doors are produced using frame joints.

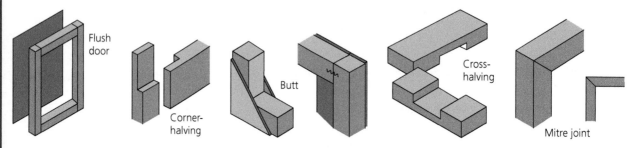

Figure 4.26 Frame joints

Adhesives

- The two most popular glues used when joining wood are PVA and contact adhesive. PVA is a very strong glue but it takes quite a long time to dry. Contact adhesive is only a medium-strength glue but, as the name suggests, it provides a quick joint.
- Epoxy resins can be used for joining wood to other materials such as metals and polymers.

Woodscrews

- Woodscrews are a quick and convenient method of fastening together two or more components.
- Modern woodscrews are made from steel but have a protective coating to stop them from rusting.
- They are designed to be used with power tools, such as a cordless drill for fast fixing.

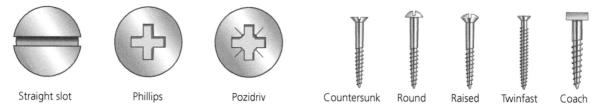

Figure 4.27 Common screwdriver slots

Figure 4.28 Common screw heads

KD fittings

- Knock-down (KD) fittings are often used with 'flat-pack furniture'.
- They enable furniture to be sold unassembled, taken home in a flat cardboard box and assembled using simple tools.

Now test yourself answers at www.hoddereducation.co.uk/myrevisionnotes

- The concept of flat-pack furniture has greatly reduced the cost of buying furniture.
- Traditionally, a family would invest in a piece of furniture that would then be handed down through the generations. Now people can afford to replace furniture as fashion and styles change.

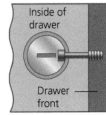

Figure 4.29 **A cam lock (KD) fitting**

Lamination and steam bending

Complex curves can be formed in wood by the processes of laminating and steam bending.

- Laminating involves gluing veneers of natural timber in between two halves of a mould.
- When steam bending timber, a solid section of natural timber is first steamed for several hours to make it pliable. It is then clamped into a mould and held for several hours until cool.

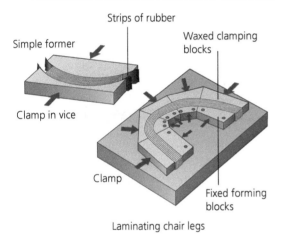

Figure 4.30 **The process of laminating**

Figure 4.31 **The process of steam bending**

Veneering

- Veneers are thin sheets of natural wood that can be applied to manufactured boards to enhance their appearance.

Computer-aided manufacture

- A laser cutter can cut thin sections of natural timber and manufactured boards from a CAD drawing.
- A laser cutter can also be used to etch an image or working onto the surface of timber.
- A 3D router can be used to produce a 3D solid timber product from a CAD drawing.
- Close-grained hardwoods, MDF and modelling foams work best with the 3D router.

Metals

Non-permanent methods of joining metal

- Nuts, bolts and washers are the most common method of non-permanently fixing metal components together.
- Machine screws allow metal components to be non-permanently fixed together.

Figure 4.32 **Nut, bolt and washer**

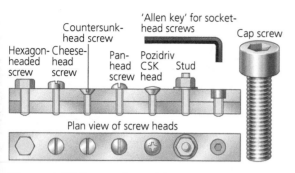

Figure 4.33 **Types of machine screw**

Permanent methods of joining metal

Riveting

- Riveting is a mechanical method of permanently fixing metal parts together.
- A hole is drilled through the metal components.
- A rivet is then placed in the hole and hammered into shape. When pop riveting, a rivet gun is used to form the rivet.

Soft soldering

- Soft soldering can be used when attaching copper pipe fittings together.
- The joint is first cleaned, a paste flux is applied and then it is heated with a torch. The solder then melts and flows into the joint.
- Solder that is to be used for plumbing is made from tin and copper.

Hard soldering

- Hard soldering is a similar method to soft soldering but is used for joining precious metals such as gold and silver.

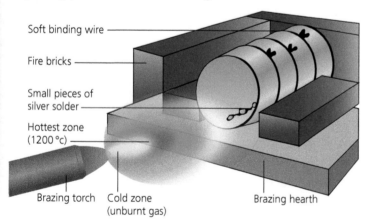

Figure 4.36 **The hard soldering process**

Brazing

- Brazing is a similar process to hard soldering but uses brass as a solder to attach steel components together.

Welding

- Welding uses heat to melt the surface of a metal. Once the metal is in a molten state it pools together and any gaps are filled with a filler rod.

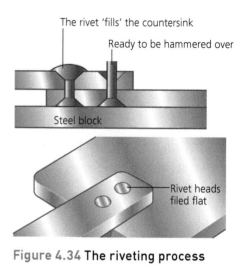

Figure 4.34 **The riveting process**

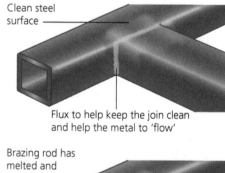

Figure 4.35 **Soldering copper piping**

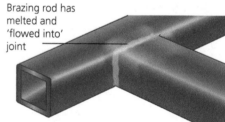

Figure 4.37 **Brazing**

Gluing

- Epoxy resins can be used to glue metals together to form a low- to medium-strength joint. Both surfaces of the metal should have a clean, keyed surface. (Keying involves lightly scratching the surface with an abrasive paper to give the adhesive something to bond to.) Epoxy resin consists of two parts, an adhesive and a hardener. These should be mixed together in equal quantities, applied to the metal surface and then clamped until the glue sets.

Machining

Centre lathe

- The centre lathe allows a cylindrical metal bar to be machined. The ends of the bar can be 'faced off', making them flat and smooth after being cut with a hacksaw.
- The diameter of the bar can be reduced by a process known as 'parallel turning' or a cone can be formed by 'taper turning'.
- The ends of metal can also be accurately drilled using a drill attached to the tailstock of a centre lathe.

Milling machines

- **Milling** machines will machine flat surfaces and allow slots and grooves to be cut into metal.

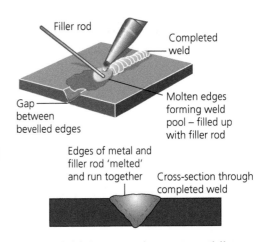

Figure 4.38 **Oxy acetylene gas welding**

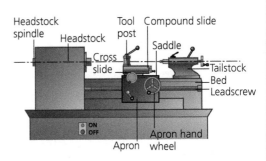

Figure 4.39 **Centre lathe**

> **Milling:** cutting grooves and slots into metal.

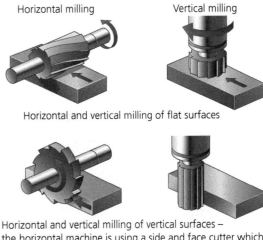

Horizontal and vertical milling of flat surfaces

Horizontal and vertical milling of vertical surfaces – the horizontal machine is using a side and face cutter which cuts on its side and on its diameter

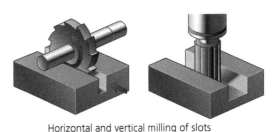

Horizontal and vertical milling of slots

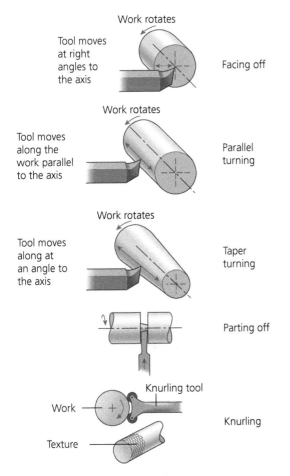

Figure 4.40 **Lathe operations**

Figure 4.41 **Milling operations**

CNC lathes and milling machines

- Metal is ideally suited to CNC machines using CNC lathes and milling machines.
- A CNC lathe can parallel turn, face off, taper turn and even cut screw threads.
- A CNC milling machine can cut slots, grooves and machine edges and create flat, smooth surfaces in metal.
- Both machines follow CAD drawings to a high level of accuracy.
- They are faster and more consistent than machining by traditional methods on a centre lathe or a milling machine.

Plasma cutter

- A plasma cutter performs similar operations to a laser cutter but can cut and engrave quite large thicknesses of metal.

Figure 4.42 **A plasma cutter**

Polymers

Joining polymers: temporary joints

- Polymers can be temporarily fastened together using nuts and bolts. Fixings designed specifically for use with polymers have larger heads to displace the pressure over a larger area.
- Self-tapping screws and panel trim fixings provide a quick and efficient way of fixing polymer components together. Look inside a car to see how the plastic panels have been fixed in place.

Figure 4.43 **Pop rivets and a rivet gun**

- Polymer sheets can be riveted together using aluminium rivets, pop rivets or bifurcated rivets. Care must be taken not to apply excessive force as the polymer sheet can be easily damaged.
- Traditional butt and flush hinges can be used with polymers. However, some polymers such as expanded polystyrene (EPS) can be formed with an integral (built-in) hinge.
- Traditional catches can be used with polymers, but integral catches can be built into the design.

Figure 4.44 **EPS food container with integral hinge**

Figure 4.45 **Food container with handle and catch**

Joining polymers: permanent joints

- Tensol cement (dichloromethane and methyl methacrylate) is a clear solvent adhesive that is formulated to glue polymers such as PMMA.
- Certain polymers can be welded together using a hot-air gun to melt the surfaces of the polymers. The seam on a plastic bag is welded together using a heated clamp.

Machining

The centre lathe and the milling machine

- A metal working centre lathe and a milling machine can perform all the same operations on polymer-based materials as they can on metal-based materials. See 'Metals' earlier in this section for more information on turning and milling operations.

Computer-aided manufacture

- Polymers are ideally suited to volume production using CAM.
- A vinyl cutter uses a computer-controlled blade to cut self-adhesive vinyl by following a computer-aided design.
- A laser cutter uses a laser beam to cut and etch into certain polymers such as PMMA following a CAD drawing.
- A 3D router will mill slots and profiles into most polymers and is most often used with modelling foams such as Foamex.
- CNC lathes and CNC milling machines will perform the same operations as centre lathes and milling machines but are computer controlled following a CAD drawing.
- A 3D printer can produce a complete 3D polymer product from a 3D CAD drawing.

Figure 4.46 **CNC lathe machining a component**

Injection moulding

Many products that we use today are produced by injection moulding, for example a chair, pen, smartphone case. The injection-moulding process is as follows:

- Polymer granules are fed into the hopper, which feeds the granules into the heating chamber.
- An Archimedean screw transports the granules along the heating chamber, which gradually turns them into a molten state.
- The molten plastic is then injected into the pre-prepared mould.
- The mould is cooled and the component is removed.

Vacuum forming and blow moulding

Both these processes use a vacuum-forming machine to produce thin-walled polymer shapes such as yoghurt cartons and clam shell packaging.

- Place the mould onto the platen and lower the platen.
- Clamp the polymer sheet onto the machine and then heat the sheet until soft.
- Blow a dome (only necessary for tall moulds).
- Raise the platen.
- Switch on the vacuum.

- Remove the heat and allow to cool.
- Remove the mould from the formed polymer sheet and trim.

Figure 4.47 **Vacuum-formed food tray**

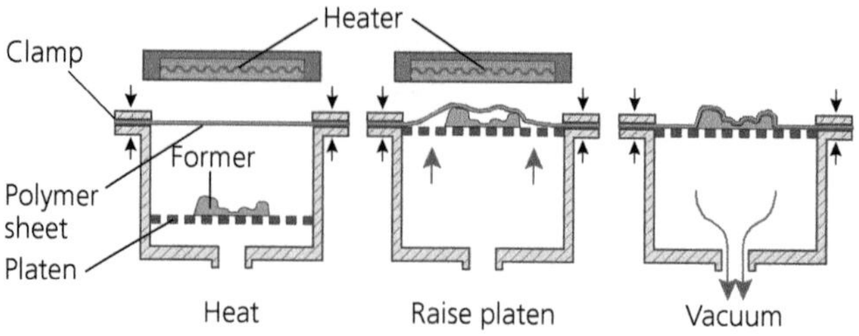

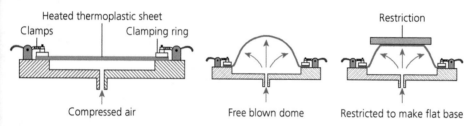

Figure 4.48 **The vacuum-forming process**

The blow moulding process

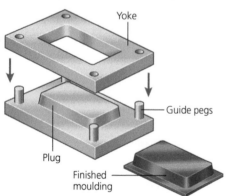

Figure 4.49 **The blow moulding process**

- Clamp the polymer sheet onto the machine.
- Heat the sheet until soft.
- Blow a dome.
- Remove the heat and allow to cool.
- Remove the blown polymer sheet and trim.

Press forming

Thicker three-dimensional shapes can be produced by **press forming**.

Figure 4.50 **The press forming process**

- An acrylic sheet is warmed in an oven until soft.
- The sheet is placed over the 'plug'.
- The 'yoke' is placed over the top of the acrylic and pressure is applied.
- The acrylic sheet will take the shape of the mould.
- The acrylic sheet is then left to cool, removed from the mould and trimmed.

Line bending

Line bending using a strip heater is an effective way of producing a straight-line bend in acrylic.

> **Typical mistake**
>
> When asked to describe a manufacturing process such as vacuum forming, students will often produce weak, inaccurate diagrams. Make sure that you can draw accurate, fully labelled diagrams of the main forming processes of vacuum forming and blow moulding.

> **Blow moulding**: a method of shaping a thermoforming polymer by blowing into a dome.
>
> **Press forming**: a method of shaping a thermoforming polymer by heating and pressing into a mould.

Now test yourself answers at **www.hoddereducation.co.uk/myrevisionnotes**

The line bending process is as follows:

- Place the acrylic sheet directly above the hot wire.
- Regularly check to see if the acrylic has softened as it heats up.
- Once softened, remove the sheet and bend to the desired angle. (A jig can help achieve the correct angle.)

Paper and boards

Folding

- Paper and thin card can be folded easily by hand.
- Scoring will help fold thicker card and ensure a clean, sharp crease.
- Foam board can be folded by cutting through the foam in one of two ways:
  - ○ Hinge cutting is where the foam board is cut partway through so that the bottom layer of card acts as a hinge and the card can be folded backwards.
  - ○ Vee cutting is where a 'V'-shaped cut is made in the foam board and the material removed. This allows the foam board to be folded inwards and gives a clean, tidy fold.
- PVC foam cannot be folded unless it is cut partway through, in a similar way to foam board.
- Corriflute can be folded by cutting a section of material away from the top layer between the flutes.
- Styrofoam cannot be folded.

Embossing and debossing

- Embossing and debossing give paper and card a three-dimensional image that can be seen and felt.
- Embossing creates a raised area on the paper or card that stands out slightly.
- Debossing has the opposite effect and creates a sunken or lowered area.

Binding methods

- Different binding methods are used for books, magazines and other large publications.
- Most publications are sewn together.
- Saddle stitching is the most common and least expensive method used for books and magazines.
- Loop stitching works in a similar way to saddle stitching but allows extra pages to be added at a later date.
- Side or stab stitching uses wire, then a section along the edge is added to cover the wire.
- Sewn binding is similar to saddle stitching but uses thread instead of wire.
- Documents with large numbers of pages are bound in stages. The pages are divided up and stitched into small sections called signatures. The signatures are then joined together using one of the methods below:
  - ○ Perfect binding is where the signatures are glued together into a wrap-around cover.

Typical mistake

Paper and cardboard are often considered to not be strong materials, but when folded or shaped in certain ways to create 'structures' they are incredibly strong for their weight.

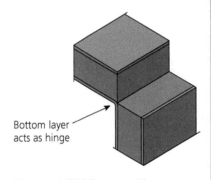

Bottom layer acts as hinge

Figure 4.51 Hinge cutting

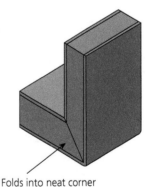

Folds into neat corner

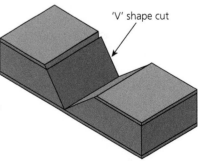

'V' shape cut

Figure 4.52 Vee cutting

○ Tape binding is similar to perfect binding but uses an adhesive tape wrapped around the signatures to hold them all in place.

○ Case binding involves gluing the signatures to end papers, which are glued to the spine of the book cover.

Other binding methods

● Ring or comb binding uses holes punched along the edges of the pages and a spiral ring or plastic comb binder holds the pages loosely together.

● A plastic spine is a U-shaped length of plastic that grips the pages placed between.

● Stud binding (also known as screw or post binding) is where holes are drilled through all the pages, then a stud is pushed through and an end cap fitted.

11 Surface treatments and finishes

Metals

● Ferrous metals require a finish to be applied, otherwise they will rust.

● Non-ferrous metals may be given a finish to enhance their appearance but precious metals such as gold and silver are usually just polished.

● Finishing always begins with surface preparation. The surface should be sanded smooth and be free from dirt, dust and oil.

Dip-coating

● Dip-coating covers the metal in a layer of polyethylene.

● The metal is fully cleaned and then placed into an oven and heated to a temperature of 200°C.

● The metal is then dipped into a fluidising bath for a few seconds.

● The polyethylene sticks and melts onto the hot surface, leaving a smooth, shiny coating.

Powder coating

● This is an industrial equivalent of dip-coating. The metal is fully cleaned and then placed into an oven and heated to a temperature of 200°C.

● The metal is sprayed with a layer of powdered polyethylene from an electrostatic spray gun, which ensures an even coating over the metal.

● The metal is then placed back into the oven to cure.

Figure 4.53 **Tools with dip-coated handles**

Galvanising

● This is an industrial process that involves coating the metal in zinc.

● The metal is fully clean and then dipped into a bath of molten zinc.

● The zinc provides a very durable and corrosion-resistant barrier.

Anodising

● Anodising is an electrolytic finishing process that is specific to aluminium.

● The aluminium is fully cleaned in a series of mechanical and chemical methods.

Figure 4.54 **Galvanised steel safety barrier beside a road**

- The aluminium is then placed into a chemical bath where a current is passed through the metal, forming an oxidised layer.
- The oxidised layer hardens the surface and provides a corrosion-resistant protective layer.
- Coloured dyes can be added to the chemical bath and colour the oxidised layer.

Enamelling

- Enamelling can be used to decorate jewellery or to protect household appliances.
- Enamelling involves coating metal with powdered glass that is then fused to the metal by heating it in a kiln to a temperature above 800°C.

Oil blacking

- Oil blacking produces a very thin layer of black oxide on the surface of ferrous metals.
- It is used on machine tools and components to prevent them from rusting.
- A ferrous metal component is heated and then soaked in an oil bath.

Painting

- Painting metals prevents corrosion and enhances their appearance. Metals can be painted any colour.
- Surface preparation is very important – metals should be free from dust, dirt and oil.
- First a primer coat is applied, followed by several layers of coloured gloss paint.
- Paint can be applied by brush or may be sprayed on to the metal surface.

Figure 4.55 **A range of anodised bike components**

Figure 4.56 **Enamel jewellery**

Typical mistake

If a question asks you for a detailed description of how to prepare and apply a finish to a metal component, make sure that you include details of any surface preparation that may be required. Also include details of any health and safety issues that may arise.

Timber

REVISED

A finish is required on most natural timbers as they need to be protected from the weather. Newly-converted planks of timber must be seasoned to remove moisture.

Surface preparation

Preparing the surface of the timber is an essential part of the finishing process.
- The timber should be sanded using a variety of grades of sandpaper to remove any marks and to smooth the surface.
- The surface should be free from dust, dirt and oil.

Types of finish

- A wood stain will change the colour of the timber but offers little protection.
- A wood preservative soaks into the timber and protects it from both moisture and insect attack. Preservatives can also include a stain. They are generally used on garden sheds and fencing.
- Tanalising is a commercial method of applying a preservative. The timber is pressure treated with the preservative. Tanalised timber is used extensively in the manufacture of patio decking.

- Varnishes offer good protection to the timber. A clear coating is normally used but they are also available in a range of colours.
- Oils, such as Danish oil and teak oil, are easy to apply to timber and give a reasonable level of protection. They are not particularly long lasting, however, and will require recoating each year.
- Wax polishes give a low level of protection and are generally used on top of a varnished surface. French polishing is a specialised method of applying polish and is used only for very high-quality furnishings.
- Paints give a high level of protection to timber and are available in a wide range of colours. Most paints require an extra level of preparation. If there are knots in the timber they must first be primed, the surface of the timber must then have an undercoat before the final gloss/satin/matt coat is applied.
- A veneer is often applied to manufactured boards to improve their appearance.

Polymers

- Polymers are self-coloured and usually have an immaculate, high-quality finish.
- They are quite resistant to wear and tear and are not affected by water or by many chemicals, so they require little finishing.

Polishing

- The cut edges of a polymer can be brought back to a high-quality finish by draw filing, sanding with 'wet or dry' paper and then polishing using a metal polish or by buffing on a buffing wheel.

Printing

- Decoration and texture can be applied to the surface of a polymer by the pad or screen-printing method.

Vinyl decals

- Self-adhesive vinyl decals can be stuck onto the surface of a polymer-based product.

Textured finishes

- A textured finish is generally created at the time of moulding. The finish would be integrated into the shape of the mould. Textured finishes are often found on polymer-based products to give grip to the otherwise slippery, shiny surface.

Figure 4.57 **Vinyl decals**

Figure 4.58 **Draw filing acrylic**

Figure 4.59 **Polishing acrylic**

Now test yourself

1. Explain why newly-felled timber must be seasoned. [4]
2. Explain why some solvent-based adhesives are unsuitable for Styrofoam and foam board. [2]
3. Name two methods of permanently joining metal together. [2]
4. Name two non-permanent methods of joining metal together. [2]
5. Use notes and sketches to describe the process of vacuum forming. [6]
6. Explain the advantages of using KD fittings to both the customer and the manufacturer of wooden furniture. [6]
7. Use notes and sketches to describe how you would dip coat a metal component. [6]
8. Use notes and sketches to describe how to produce a bend in wood using the steam-bending technique. [6]
9. Explain why many polymers do not need an applied finish. [3]
10. Explain the purpose of a pilot hole. [2]
11. Give two reasons why it is necessary to apply a finish to patio decking. [2]
12. Use notes and sketches to describe the life cycle of a steel can. [6]
13. Describe some of the reasons why food packaging cannot always be recycled. [4]

Exam practice

1. Explain the advantages of using manufactured boards rather than natural timber. [4]
2. Complete the following sentence: Nitinol is one of the most common shape memory alloys made up of .... and .... [2]
3. Describe what happens to the fibres in paper each time it is recycled and how this affects the recycled paper's properties and uses. [3]
4. Describe the life cycle of a soft drinks can made from aluminium. [6]
5. Study the picture of rainwater guttering.

   (a) Name a suitable thermoforming polymer for the rainwater guttering. [1]
   (b) The rainwater guttering has been made by extrusion. Explain why extrusion is a suitable production method for the manufacture of rainwater guttering. [3]
6. Explain the process of laminating and why this is done to certain paper and card products. [3]

5 Core skills

1 Understanding design and technology practice

Context of a design

REVISED

The **context** of a design incorporates many things such as:
- ○ the surroundings or environment where it will be used
- ○ the different **users** and **stakeholders**
- ○ the purpose of the end product
- ○ social, cultural, moral and environmental considerations.

- A product that is designed properly within context will fulfil its purpose exactly, giving the users and stakeholders what they require with a minimum of interaction or inconvenience.
- If design takes place without consideration of the context it will result in a final product that does not fully meet the needs of the users or stakeholders.

> **Context:** the settings or surroundings in which the final product will be used.
>
> **User:** the person or group of people a product is designed for.
>
> **Stakeholder:** a person other than the main user who comes into contact with or has an interest in the product.

Content map/task analysis

REVISED

A useful method of considering the context of a design is to create a context map or task analysis showing all the possible factors that could or should influence the design, such as:

- **Who** are the primary users, secondary users and other stakeholders? For example, their age, gender, physical mobility.
- **Why** is there a need for this product? What is the problem? What are the restrictions on the design?
- **Where** will the product be used? What type of environment will it be used in? For example, is it for use indoors or outdoors?
- **What** must the product do? What is its primary function? Does it have just one main purpose or must it fulfil a number of different requirements?
- **When** is the product used? Is the product used at certain times of day/ night? Is it used at certain times of the year then stored for long periods of time between uses?
- **How** is the product meant to function? How will it be stored, transported, maintained? For example, will it have to function without making any noise?

Many of these considerations will also overlap and influence each other.

> **Exam tip**
>
> Toys are designed for children, but parents purchase the toys and have to store, carry, clean and maintain them. Although the child is the main user, the parents are a major stakeholder and the designer must also consider their needs.

Figure 5.1 The WOLF rechargeable torch was designed for use in extremely cold and wet environments

2 Understanding user needs

Needs and wants of the end user

REVISED ☐

- One of the main considerations for any design must be the user group and their needs.
- The main person or group of people who will be using the product are known as the **primary users**.
- Assessing user and stakeholder needs will require data to be collected.

Primary data

- **Primary data** is collected first-hand from the main user or stakeholders, for example questionnaires, surveys, interviews or studies you conduct yourself.
- If a product is to be marketed and aimed at a wide user group then hundreds of questionnaires, interviews and tests across the spectrum of different users will need to be conducted.
- Failure to carry out sufficient research can lead to inaccurate primary data and a product that does not fulfil all of the users' or stakeholders' needs.

Primary user: the main user of the product.

Primary data: research collected 'first hand' by yourself.

Secondary data: research collected by others.

Collecting primary data

- Surveys and questionnaires are an effective way of gathering information from people, providing the questions are well thought out and worded.
- Questions can be open or multiple choice depending on the type of information you require.
- Limiting the answers to a choice of three or four options means that you won't get irrelevant answers.
- Finding people within your intended user groups to fill out surveys is of prime importance but can be difficult.
- People who fill out your survey or questionnaire must fall into the same user group as your intended users or the information collected will be inaccurate and will result in a flawed final product.
- Many free online survey programs are now available, meaning you can email people who fit into your intended user group.

Secondary data

- **Secondary data** is 'second-hand' data which has already been collected by someone else.
- It is usually much easier to find than primary data but it may already be out of date.
- Secondary data can be gathered from a range of sources, such as websites, books, test reports and journals.

- Using secondary data can save time as it is much quicker than carrying out testing, interviews and questionnaires and is therefore less expensive.
- Data collected is not as accurate as primary data because it is not specific to the designer's or user's exact needs.

Collecting secondary data

- Images and information on existing products from the internet are based on the reviews of others and/or the manufacturer's claims.
- It is better to find a 'real-life' existing product which you can handle, use and test yourself.

Now test yourself

TESTED ☐

1. State three ways of collecting primary data. [3]
2. Explain the advantages and disadvantages of using secondary data for research. [4]

---

3 Writing a design brief and specifications

Design briefs

REVISED ☐

- A design brief is a concise description of the task that the designer will undertake to solve the design problem or achieve what the **client** wants.
- The design brief can be set by the client or decided in discussion between the client and the designer.
- The brief should be short and give a clear outline of what the required results of the design are.
- The designer must refer to the brief during the design process to ensure they are working towards achieving this.

Writing a design brief

- Before starting it is important that the full extent of the problem and context has been fully explored.
- What a client wants and what they might actually need are not always the same thing.
- When the brief focuses more on the product than the problem, this can lead to **design fixation** and prevent alternative approaches being considered.

Specifications

REVISED ☐

- The specification is a set of requirements that the product must meet or constraints that it must fit into.
- Research carried out into the problem should also influence what goes into the specification.
- **Open specifications** state **criteria** that the product must meet but do not specify how this must be achieved.
- **Closed specifications** are more detailed and state what must be achieved and how certain criteria must be met, for example by specifying the tools, materials or processes that must be used.

> **Client:** the person the designer is working for (this may or may not be the user).
>
> **Design fixation:** when a designer limits their creativity by only exploring one avenue of design or relying too heavily on features of existing designs.
>
> **Open specification:** list of criteria the product must meet but not specifying how it must be achieved.
>
> **Criteria:** specific goals that a product must achieve in order to be successful.
>
> **Closed specification:** list of criteria stating what must be achieved and how it must be met.

Now test yourself answers at www.hoddereducation.co.uk/myrevisionnotes

Writing a specification

The specification will usually cover areas such as:

- the primary and secondary functions that the product must achieve (what it must do)
- any specific requirements of the users/stakeholders
- materials and components that must be used or avoided
- maximum or minimum dimensions, weights and size constraints
- financial constraints (how much it should cost to produce)
- aesthetic factors (how the product must look or feel)
- anthropometric and ergonomic requirements
- environmental standards or constraints
- safety features and restrictions
- relevant manufacturing standards
- legal requirements
- how long it should be expected to last for.

Various websites and software packages are available to assist when writing a specification by providing prompts or questions about the intended product, for example ACCESS FM, SCAMPER.

Now test yourself

TESTED

1 What are the key differences between the brief and the specification? [2]
2 What can help prevent design fixation? [2]

4 Investigating environmental, social and economic challenges

Environmental considerations

REVISED

- Until recently, manufacturers produced products as cheaply and quickly as possible with little concern about the environmental impact.
- Social pressures to have the newest products and the influences of fashion and trends mean people throw away rather than repair things. This is known as a throwaway society.
- The **throwaway society** follows a **linear economy** – see Section 1, Topic 2. Once the product is broken or no longer wanted it is simply disposed of.
- A **circular economy** uses as few resources as possible and extracts the maximum from them, using them for as long as possible – see Section 1, Topic 2.

Figure 5.2 **The use of plastic when designing products can have harmful effects on the environment**

Throwaway society: a society that excessively consumes and wastes resources.

Linear economy: raw materials are used to make a product; waste is thrown away.

Circular economy: extracting the maximum value from resources which are then kept in use as long as possible, recovered and regenerated into new products instead of thrown away.

Designing environmentally friendly products

Eco-design is designing sustainable products that will not harm the environment by, for example:

- choosing materials which are sustainable, recyclable and non-toxic and do not require as much energy to process
- designing products that:
  - are fuel and material efficient
  - last as long as possible so there is less replacement of parts
  - function to the best of their potential
  - can be fully recycled
  - use local materials, resources and labour

A simple set of rules that cover the above are the Six Rs – see Section 1, Topic 5.

Social challenges

Positive and negative impacts of products

- Products can have a negative impact on society that the designer may not have foreseen.
- While it is not always easy to predict what the effects of a product might be, designers must try to consider the possible negative implications and decide whether the product is worthwhile.

Cultural awareness

- **Cultural awareness** is considering how different cultural groups may be affected by a product. Different **cultures** may view products differently depending on their beliefs, ideas and experience.
- A product designed for a particular market sector (for example the UK market) should not be offensive or upsetting to anyone.
- Designing new products that continue traditional processes, skills and materials can help preserve cultural identity.

Economic challenges

- The manufacturing processes, transportation, use and disposal of a product can affect the economic needs of people.
- The manufacture of a new product may create jobs, boosting the overall economy of the area.
- If the product is manufactured by modern CAM machines instead of by humans, it could lead to the loss of jobs and in turn affect the economy in that area.
- Many products are made abroad because labour costs are much cheaper. Some workers in these countries are poorly paid, work in dangerous or unhealthy conditions and work long hours without breaks or holidays.
- In some countries child labour is used.
- Designers should make sure their clients follow the guidelines for fair treatment of workers set out by organisations such as the European Institute for Crime Prevention and Control, the GLA (Gangmasters and Labour Abuse Authority) and AWARE (the Alliance for Workers Against Repression Everywhere).
- Manufacturers of Fairtrade products ensure that their workers are treated fairly. For more on Fairtrade, see Section 1, Topic 5.

> **Eco-design:** designing sustainable products that will not harm the environment by considering the effects of the technology, processes and materials used.
>
> **Cultural awareness:** understanding the differences in attitudes and values between people from other countries or other backgrounds.
>
> **Culture:** the ideas, customs and social behaviours of a particular people or society.

Anthropometrical and ergonomic issues

- **Anthropometrics** uses statistics and measurements of various parts of the human body taken from people of different ages, genders and races.
- This is used to determine the sizes, shapes and forms of products so that they meet the needs of the intended user.
- In some cases products are designed to fit the average person, in others they may be designed to fit the largest or smallest.
- **Ergonomics** is the relationship between people and the products they use.
- Ergonomic data is used to make products more comfortable and easy to use by considering the shape of the human body and the force a person can apply to something.
- Ergonomics and anthropometrics are used together when designing products to ensure the outcome not only 'fits' the intended user but is as comfortable and easy to use as possible.

> **Anthropometrics:** the study of the sizes of people in relation to products.
>
> **Ergonomics:** the relationship between people and the products they use.

Now test yourself

TESTED

1. What is the difference between anthropometrics and ergonomics? [2]
2. Describe how mobile phones have had both a positive and a negative impact on our society. [8]

5 Developing ideas

Testing and evaluating ideas

REVISED

- Innovative design is exploring new ways of doing things or trying to approach problems from different angles or perspectives.
- To get different perspectives on a problem in order to explore new ideas, designers can use different sources of information:
  - Focus groups: a group made up of different people who each give their views and experiences of a problem or discuss their perceptions and attitudes towards an existing product or idea.
  - Existing solutions: by looking at existing solutions to a problem or products designed to solve similar problems and considering which aspects work well or need improvement.
  - Biomimicry: by looking at different aspects of the natural world such as naturally occurring structures and features, then considering how these methods, shapes or forms could be incorporated into the design (for more on biomimicry, see Section 1, Topic 4.
- **Development** is the process of selecting ideas, elements, materials and manufacturing techniques from initial ideas and using them to explore and produce better designs or ideas.
- Development is the part of the design process where designers try out new things and make creative decisions based on what works and what doesn't.
- **Modelling** allows designers to try out and test ideas or parts of designs by making scale models. Modelling and testing allow designers to see whether a design will work and then make further decisions or developments.

> **Development:** the creative process of selecting ideas, elements, materials and manufacturing techniques from initial ideas and using them in new ways to explore and produce newer and better designs or ideas.
>
> **Modelling:** trying out and testing ideas or parts of designs by making scale models.

- Models and ideas that do not work are just as valid since often more can be learned from a failed or unsuccessful model.
- By taking risks and trying new things, designers can come up with innovative ideas. Each step along the way is a vital part of the development process.

Critical analysis and evaluation

- Critical analysis and evaluation go hand in hand with development but can be done at any point in the design process.
- Critical analysis will assess the suitability of the design against given criteria to check it meets the necessary requirements.
- The list of criteria will depend on the product but will cover things such as:
  - Product function: Does the product do what it is supposed to? How well does it do this? Is it easy to use? Is it comfortable to use?
  - Aesthetics of the product: Does it look appealing? Does it feel nice?
  - Anthropometrics and ergonomics: Is it the correct height, length, diameter, etc.? Does it suit all users? Can it be adjusted to suit different users? Will it fit where it is supposed to?
  - Cost: Will the materials be expensive to purchase? Will the manufacturing costs be acceptable? Will the final price of the product be acceptable?
  - Materials: Are the materials used of high quality? Are the materials suitable for the product? Are the materials easy to source/readily available?
  - Construction methods: Is it well made? Are the skills/processes required to make it readily available? Is it quick to manufacture?
  - Health and safety: Is it safe to use? Does it meet relevant health and safety standards?
  - Environmental considerations: Are the materials from a sustainable source? Can the product be easily disassembled? Can the materials be recycled? Are the processes used harmful to the environment?

Refinement and modification

- Following the critical analysis, the designer can make further changes or modifications in order to address any areas where the design is not meeting the criteria or not performing well.
- The modified design is then re-analysed and if further improvements are still needed, the design will be modified again and the same process repeated.
- This cycle of refinement, analysis and re-designing is called **iterative design**.
- Each new 'iteration' of the design should be a slightly improved version of the last until the final iteration, which should meet all the criteria as much as possible and is the best possible solution.

> **Iterative design:** a repeated cycle of quickly implementing designs or prototypes, gathering feedback and refining the design.

Now test yourself

1 State four criteria that would be considered when critically analysing a product. [4]
2 Describe an example of how a designer might use biomimicry. [3]

6 Using design strategies

Design strategies

Collaboration

- Many designers work in **collaboration**, in pairs or in groups. Usually a design business will employ a number of designers who work together on specific design projects.
- By discussing, sharing and working with each other they bounce ideas back and forth, refine elements or go down completely new avenues of investigation.
- Once a designer has created some initial ideas or concepts, a focus group comprising the client, user and main stakeholders gives feedback on the designs.

User-centred design

- **User-centred design (USD)** puts the user at the 'centre' of the design process.
- The USD process has four main stages:
  - Specify the context of use: identify the users of the product, what they will use it for and under what conditions it will be used.
  - Specify requirements: identify any user goals that must be met for the product to be successful.
  - Create design solutions: this may be done in stages, building from a rough concept to a complete design.
  - Evaluate designs: evaluation is done through usability testing with actual users.

Systems thinking

- **Systems thinking** is when you consider the product you are designing as part of a larger system or experience.
- The opening of the packaging, maintenance of the product, use of the product and disposal or exchange of the product are all part of the experience of owning a product.
- Systems thinking considers the whole problem and how to provide the best service to the user.

> **Collaboration:** a number of designers working together on specific design projects.
>
> **User-centred design (USD):** looking at and checking the needs, wants and requirements of the user at every stage of the design process.
>
> **Systems thinking:** considering a design problem as a whole experience for the user.

Now test yourself

1 Describe the benefits and possible drawbacks of collaboration when designing. [4]
2 Name the four main stages of user-centred design. [4]

7 Communicating ideas

Formal and informal 2D and 3D drawings

- Two-dimensional drawings are useful for showing simple profiles (shapes), the layout of parts within a design, or cross-sections through a design.
- Three-dimensional drawings are useful for showing design ideas.

Freehand sketching

- Freehand sketching refers to sketches made without the use of templates, grids or other drawing aids.
- Feint lines may be sketched to 'crate out' simple block shapes which the designer will later refine into the required shape.
- Drawings can be 2D or any style of 3D.
- Brief annotation may be used.

Oblique drawing

- An **oblique drawing** is a basic, low-skilled form of three-dimensional drawing.
- A two-dimensional profile is projected into a 3D shape by drawing lines at 45°.
- A 45° set square is usually used, for accuracy.

Isometric drawing

- This is a 3D drawing method that produces a more accurate representation of a shape.
- An angle of 30° is used for the projected lines.
- A 30° set square or isometric grid paper is often used, for accuracy.
- All lines are drawn to their full length.
- Large isometric drawings can appear distorted in shape because they have no 'perspective'.

Perspective drawing

- A **perspective drawing** is a 3D drawing method that achieves a very realistic representation.
- In two-point perspective, two 'vanishing points' are drawn at either end of an imaginary horizon.
- The projected lines from the front edges of the drawing all move towards the vanishing points, causing them to converge.

Thick and thin line technique

- This is a simple method of improving the impact of a 3D sketch.
- A thicker, bolder line (using a pen or a dark pencil) is used to 'go over' all edges of the drawing where the detail on an adjacent edge is not visible.

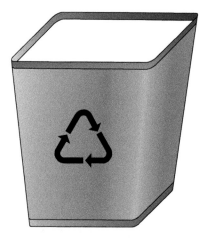

Figure 5.3 Oblique projection drawing

Exam tip

Questions may ask you to draw a 3D (isometric) projection of an object from a series of 2D (orthographic) drawings, or vice versa.

Oblique drawing: a basic 3D drawing method, using lines projected at 45°.

Perspective drawing: a realistic 3D drawing method.

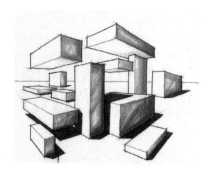

Figure 5.4 Two-point perspective drawing

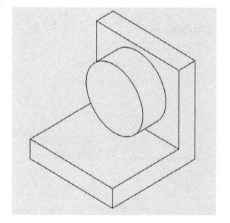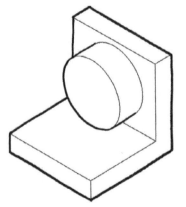

Figure 5.5 Thick and thin lines used on an isometric drawing

Colour and shadow

- This is a skilled technique of showing surface patterns, texture and depth.
- It is essential for representing textiles and fabrics.
- Shadows and shading can show depth and shape of surface.
- Different **rendering** techniques can be used.
- 3D CAD programs can produce very realistic rendered drawings.

System and schematic diagrams

- These diagrams are often used in electronic systems.
- A **system diagram** (or block diagram) is used to show the functional subsystems, how they are interconnected and the signals flowing between them.
- System diagrams are broadly divided into input, process and output subsystems.
- A **schematic diagram** (often called a circuit diagram) indicates the connections between individual electronic components and the values of the components.

> **Rendering**: the method of using colour and shading to represent the nature of a surface.
>
> **System diagram**: shows the interconnections between subsystems in an electronic system.
>
> **Schematic diagram**: a circuit diagram showing the connections between individual components.

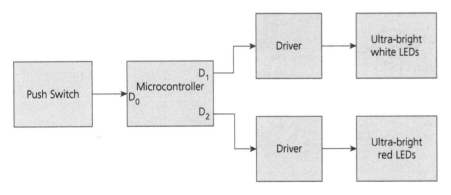

Figure 5.6 **Electronic system diagram**

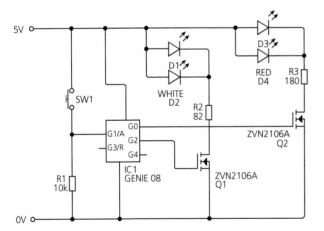

Figure 5.7 **Electronic schematic diagram**

> **Exam tip**
>
> Learn to recognise and to draw the basic circuit symbols used in schematic circuit diagrams.

Annotated sketches

Annotations should be used to add information which is not obvious, for example:

- arrows to indicate movement
- material used

- surface finish
- construction or manufacturing method
- function
- reference to commonly understood phrases, such as 'headphone jack', or 'Velcro attachment'
- hidden detail, for example 'battery compartment underneath'
- reference to weight or balance
- samples of swatches or fabric for textiles.

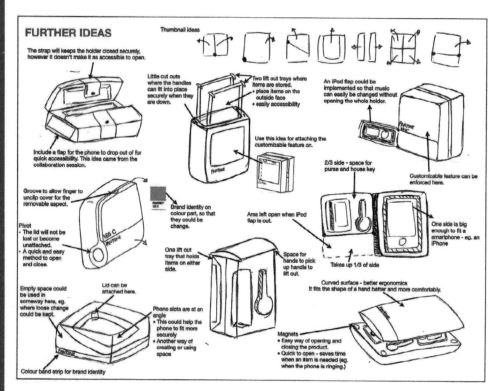

Figure 5.8 **Annotated sketching**

Exploded drawing

- An **exploded drawing** is a 3D drawing method that is used to show how components in an assembly fit together.
- **Isometric drawing** is usually used.
- The parts are separated along the 30° isometric lines, with arrows or dotted lines used to show how the parts fit back together.
- CAD programs can produce exploded drawings quite simply.

> **Exploded drawing:** a 3D drawing method to show how parts in an assembly fit together.
>
> **Isometric drawing:** a 3D drawing method, using 30° angles for projections of depth.

Fashion illustrations or fashion drawings

- Fashion illustrations or fashion drawings can be presented as a CAD drawing or may be hand-drawn.
- Some fashion drawings convey a design concept – more of a free-hand creative expression than a detailed drawing of the product.
- Fashion illustrations often exaggerate the length of the figure, particularly the legs, creating a more dramatic effect.
- Fashion illustrations and drawings can be presented in any media – freedom of expression is important to convey details such as how potential fabrics drape and fall. Fabric swatches and trims are often presented alongside an illustration.

Models can be used to produce a full-size or scaled-down version of the entire concept, or they can be used to test a small part of the design, such as the function of a component or the finish on a material.

Cardboard models

- Cardboard is easily cut and low cost.
- It can be scored, bent and joined with a variety of adhesives.
- The models are rigid and available in various thicknesses.
- They can be laser cut from a CAD drawing.
- These models are useful for testing two-dimensional linkages and mechanisms.

Foam models

- Foam core (or foam board) is similar to cardboard but more rigid. It is faced with white or black paper, giving it a very smooth surface, which can be rendered to give a realistic appearance.
- Polyurethane foam is used to create solid, 3D models which can be interacted with, for example held in the user's hands. It is available in a variety of densities which can be easily cut and shaped using hand tools or CNC machines.
- Foam models can be finished and sprayed to look very realistic.

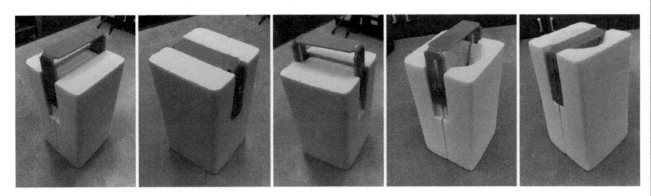

Figure 5.9 **Using blue foam to model a retracting handle concept**

Toiles

- A toile is a full-size textile model, manufactured from a cheaper fabric.
- It allows the designer to check garment fit or design and determine the position of zips, buttons, etc.

Electronic circuits

Breadboards can be used to temporarily build a circuit for testing:

- No soldering is required.
- They can be used to check component values, such as resistors.
- They can be used to test the function of a microcontroller program.

> **Breadboard:** a temporary method of constructing electronic circuits for testing.

Presentations

REVISED

A designer may need to present design ideas to a third party. This could be done using:

- digital presentations, including slides, images, videos, sound and animations
- large-format presentation boards mounted on a display, showing sketches, diagrams and rendered images of the product.

> **Exam tip**
>
> Questions may ask you to draw a flowchart to explain a manufacturing process (e.g. vacuum forming), including quality checks using decision boxes.

Written notes

- Written notes are used to formalise designing thinking and to explain design decisions.
- They are useful for explaining the creative process to a third party.

Flowcharts

- A flowchart is a graphical representation of a process.
- Different symbols are used to represent different actions in the flowchart – see Figure 2.6 in Section 2, Topic 3.
- Flowcharts are often used to explain a process, such as a sequence of manufacturing steps.
- In a manufacturing flowchart, the decisions are often quality-control checks.
- Flowcharts can be used to show a microcontroller program.

> **Typical mistakes**
>
> Take care not to confuse a manufacturing process flowchart with a microcontroller program flowchart.
>
> Understand the differences between a system diagram, a schematic diagram and a flowchart.

Working drawings

REVISED

These are formal drawings that contain enough detailed information to allow a third party to accurately manufacture the design. Working drawings are sometimes called engineering drawings.

- They are often 2D orthographic drawings, showing a front, side and plan (top) view.
- They may include cross-sections or views of hidden detail.
- They contain information such as dimensions, scale, materials and details about component parameters (values).
- They may include a parts list.
- They should be drawn using British Standards symbols and methods.
- A textile working drawing is sometimes called a 'flat'.

Schedules

- A schedule is a diagram showing how a process can be completed within a specific timeframe in order to meet a deadline.
- A Gantt chart is a common way of showing a schedule.
- In industry, schedules can be used to plan when a particular machine will be needed, or to determine when materials need to be delivered.

Audio and visual recordings

These can be used to record and present:

- findings from a focus group
- users interacting with a prototype design
- feedback and opinions about a model or a design
- evidence of the design functioning in its intended situation.

Now test yourself answers at www.hoddereducation.co.uk/myrevisionnotes

Mathematical modelling

This type of modelling is especially useful when using CAD software. Mathematical modelling can:

- simulate the effect of forces applied to the design, for example a structure
- test how a component responds to and conducts heat
- test designs in a virtual environment, saving time and expense.

Computer-based tools

- CAD software is usually used to develop and refine initial ideas.
- Edits can be made quickly, and it is easy to undo changes.
- 2D CAD designs can output to CNC machines, such as a laser cutter.
- 3D CAD models can represent the colour and texture of the materials used.
- CAD software can render the design to give it a photorealistic look.
- Common components, for example fasteners or hinges, can be selected from a parts library.

Now test yourself
TESTED

1 Name three types of 3D drawing techniques used by designers. [3]
2 Explain the difference between a system diagram and a schematic diagram in an electronic design. [3]
3 Give three benefits to a designer of making a model during the development of a design. [3]
4 Draw the flowchart symbols for:
  (i) input/output
  (ii) decision
  (iii) process. [3]
5 Describe two benefits to a designer of using CAD software. [4]

8 Developing a prototype

Prototypes
REVISED

- Prototyping involves making a one-off version of the whole product or a specific part of the design.
- **Prototypes** can be used to test parts of the design, find out the users' views and identify any problems.
- Prototypes can show up potentially fatal flaws in a design which can be addressed before the product goes into full-scale production.

> **Prototype:** an early model of a product or part of a product to see how something will look or function.
>
> **Low-fidelity prototype:** a quick prototype that gives a basic idea of a product's look or functions.

Low-fidelity prototypes

- **Low-fidelity prototypes** are produced early in the design process.
- They can be basic models of how a product will look or scale models of a part (such as a mechanism or feature) to illustrate or test how it works.

- Prototype models may be made from simpler materials such as paper or card instead of sheet metal and from plasticine or modelling clay instead of polymer or plastic.
- Low-fidelity models are cheap and quick to make and allow designers to test crucial elements of their design and achieve quick results.

High-fidelity prototypes

- **High-fidelity prototypes** are made once a design has been developed considerably or a final design has been decided.
- They will look, feel and function as much like the finished product as possible and be made using the same materials and processes as far as is feasible.
- High-fidelity prototypes take much longer to produce and are more expensive but give a more realistic idea of what the finished product will be like.

When making prototypes to develop your ideas, remember the following:

- Making something instead of just drawing it will help you to see your idea in a different way and find how you can improve it further.
- Don't spend a long time building a prototype as this will slow down the thought process and you will be less likely to change something if you have spent hours making it.
- Don't forget what the prototype is supposed to be testing and try to let the user test the product if possible.
- Don't be afraid to fail! If the prototype does not do what you want, use this knowledge to change or develop your design.

Figure 5.10 **Low-fidelity prototype**

> **High-fidelity prototype:** a detailed and very accurate prototype similar to the final product.

Aesthetics

REVISED

- Aesthetics is the way something (such as a product) is perceived by someone, based on how it looks, feels, sounds, smells or tastes.
- When designing a product, designers try to consider how people will perceive the product and aim to make it aesthetically pleasing to as many people as possible.
- Aesthetics affect attitudes to a product – owning something that looks, feels or smells good creates a positive feeling and makes products more valued, for example particular brand images.

Table 5.1 **Aspects of aesthetics**

Type of aesthetics	Factors to consider
Visual aesthetics First impressions of a product, whether the eye is drawn to it or not	Shape
	Form
	Colour
	Pattern
	Proportion
	Symmetry

Type of aesthetics	Factors to consider
Material aesthetics How a product feels when it is touched or handled	Texture Comfort Weight Temperature Vibration Shape
Sound aesthetics The sound something makes, e.g. the alarm on a clock or tones used in a mobile phone	Melody Pitch Beat Repetition Pattern Noise
Taste and smell Major considerations when designing household products and food	Strength Sweetness Sourness Texture (taste only)

Marketability

REVISED

- Most products are designed for a wide audience.
- The marketability of a product is whether it will appeal to buyers and sell sufficiently at the price to make a profit. Designers and manufacturers need to know how marketable a product is to decide whether it is worth launching.
- Prototype products are evaluated using a number of testing methods (see Section 5, Topic 9) to see whether there is enough demand for the product.
- After the evaluation, if the product is not marketable enough, it may be further developed or abandoned.

Innovative features

- Advances in technology allow for new and innovative products to be developed.
- For example, the development of smart materials has led to new products such as self-cleaning glass.
- Developments such as Bluetooth and automatic speech recognition (ASR) have added enhanced functionality to electronic devices.
- Smart fabrics have a number of uses in clothing and medicine (see Section 3, Topic 4).

Now test yourself

TESTED

1 Explain the term 'prototype'. [2]
2 Explain the difference between low-fidelity and high-fidelity prototypes. [4]

9 Making decisions

- Designers make decisions throughout the design process.
- Prototyping and asking for user feedback will test aspects of a design and can help the designer make important decisions.
- The design brief and specification should be the reference points for all decisions and the designer should constantly refer back to these.
- The most effective and useful feedback about a prototype is from the intended user.

User testing

- **User testing** is watching a user interact and use your product for its intended purpose to see how well it works, how easy it is to use and whether they actually like it.

> **User testing:** testing by observing a user interact with and use your product for its intended purpose.

Focus groups

- A **focus group** allows people to ask questions and state how they would like the product to be improved.
- This allows a wide range of responses to be attained, but opinions may be different and contradictory because of the different users' needs.

A/B testing

- **A/B testing** is used to choose between two different design ideas.
- The results show which design achieved the task the quickest or most efficiently.
- This type of testing is often used to compare a new version of a product to the existing one to see whether it works better.

> **Focus group:** group of people used to check a product design is on track.
>
> **A/B testing:** user testing to choose between two different design ideas.

Surveys and questionnaires

- Surveys and questionnaires are an easy way to gather information.
- The survey must ask questions that will provide accurate information about the product.
- A questionnaire can reveal how well the product meets users' needs, allowing the designer to make informed decisions about future modifications.
- The different types of feedback provide qualitative and quantitative data:
  - **Quantitative data** refers to data which gives specific counts and values in numerical terms such as measurements like height, weight, size, humidity, speed and age. Data can be gathered from surveys, experiments or observations and presented in the form of charts, graphs, tables, etc.
  - **Qualitative data** cannot be specifically measured but is observable by appearance, taste, feel, texture, gender, nationality, etc. It can be collected from observations, focus groups, interviews and archive material. The data is presented as spoken or written words rather than numbers.

> **Quantitative data:** specific measurable data given in numerical form.
>
> **Qualitative data:** observations and opinions about a product.

- All data gathered should be used by the designer to:
  - evaluate the performance and suitability of the product
  - make decisions about what needs to be improved
  - implement the necessary modifications and changes to the design
  - re-test and evaluate to check the effectiveness of the changes.

Now test yourself answers at www.hoddereducation.co.uk/myrevisionnotes

Now test yourself

1 State three ways of presenting quantitative data. [3]
2 State two ways of gathering qualitative data. [2]

Exam practice

1 (a) Give one advantage of using perspective drawing rather than isometric drawing when
sketching design ideas. [1]
   (b) Designers often produce three-dimensional models and CAD models when developing
ideas. Describe one advantage of using three-dimensional models rather than CAD models. [2]
   (c) Explain what is meant by a working drawing of a product. [3]
2 Designers have a responsibility to develop products that are as sustainable as possible.
Explain three ways that designers can make products more sustainable. [6]
3 Many products are designed with planned or 'built in' obsolescence.
   (a) Define the term 'planned obsolescence'. [2]
   (b) Using an example, explain why products are manufactured with planned obsolescence. [4]
4 Symbols are often used on products or their packaging. Name the symbol shown below
and explain its meaning. [3 marks]

Success in the examination

When will the exam be completed?

You will take the exam in the summer exam period of your final year, which will usually be in May or June.

How long will I have to complete the exam?

- The exam is two hours long.
- You should practise working past papers and sample questions within the allotted time.

What type of questions will appear in the exam paper?

The paper will consist of six questions broken down into smaller part questions. There will be a variety of short answer, structured and extended writing questions that will test your core knowledge and understanding as well as in-depth knowledge and understanding of your chosen endorsed route.

Low-tariff questions that rely on recall of knowledge		
Give, state, name, underline	1+ mark	These questions need a simple statement, a short phrase, a tick or an underline.
Higher-tariff questions that are more challenging, designed to test knowledge and understanding		
Describe, outline, explain, justify	2+ marks	These questions ask you to describe something in detail. The answer will be in sentences and/or in a list. Detail and elaboration are required in the answer.
		These questions may also ask you to use notes and sketches; this means that a clearly **labelled sketch or diagram** will gain marks.
High-tariff questions that are designed to test, stretch and challenge the more able learner; extended writing will be required and QWC may be assessed		
Evaluate	5+ marks	**Evaluation** involves assessing or appraising a situation, product or material, giving reasons to support answers.
Analyse		**Analysis** means examining and dissecting a situation or product, giving thoughtful, appropriate reasons to support answers. A logical chain of reasoning may be required.

Now test yourself answers at **www.hoddereducation.co.uk/myrevisionnotes**

Sample examination questions

Core knowledge and understanding

Sample question

Study the pictures below which show different energy sources – coal (Source A) and wind (Source B).

(a) Complete the table below by stating which energy source reflects the statement. [2]

Statement	Source A or Source B
This is a renewable energy source.	
This is a finite energy source.	

(b) Discuss the environmental advantages that energy source B has over energy source A. [4]

ONLINE

Candidate response

(a)

Statement	Source A or Source B
This is a renewable energy source.	*Source B*
This is a finite energy source.	*Source A*

(b) Wind is a cleaner energy source that does not create pollution when converted to electricity, unlike coal which releases pollutants into the atmosphere when burned during the conversion process. Wind energy is better for the environment as it is freely available and sustainable, and will not add to further depletion of the world's resources.

One mark for each correct response – in this case both answers are correct.

The environmental advantages of wind energy need to be fully discussed, including a comparison to coal.

Advantages have been discussed briefly and comparisons to coal have been included. To gain extra marks, the candidate could have further explained that pollution from burning coal includes releasing poisonous gases such as CO_2 into the atmosphere, adding to global warming. This adds the required level of detail. The second advantage is also correct but has not been fully explained. Further reference to when resources such as fossil fuels (including coal) run out, they cannot be replaced could have been included. Alternatively, mention could have been made that the mining of coal also has a negative impact on the environment, unlike installing wind turbines.

In-depth knowledge and understanding

Engineering design

Sample question

Electronic circuits can be constructed by different methods depending on the application.

Figure A shows a simple circuit built on a prototype board. Figure B shows a printed circuit board, manufactured by batch production for use in an electronic product.

25 mm

40 mm

Figure A Prototype board

Figure B Printed circuit board

Study the photographs and describe **three** ways in which the printed circuit board has been designed to make it suitable for batch production for use in an electronic product. [6]

Candidate response

The components are so small that they can't be soldered by hand. They have been put on the PCB by a robot which works quickly and doesn't make mistakes, so that means that a lot of PCBs can be made very quickly in a batch. The PCB has got a lot of components packed in a small space. This means you can make complicated circuits really small, which is good for handheld products. The white writing on the PCB tells the robot machine worker where to put all the parts so they don't waste any.

The candidate's response would score 4 marks. The first comment about soldering, while correct, has not been explained in terms of why this helps the PCB to be batch produced and would not score any marks. The comment relating to the components being put on by a robot is worth 1 mark, and the second mark is for explaining that this speeds up manufacture for batch production. One mark is awarded for the comment about a lot of components being packed in a small space, and another mark for saying that this is good for handheld products. The final comment about the white writing is incorrect and scores no marks.

Better use of technical vocabulary (e.g. 'surface mount technology', 'pick-and-place machines') would have improved the quality of this response.

Mark scheme

There are three points to be made: 1 mark for a valid point, 1 mark for explaining why that point makes the PCB suitable for batch production, or for use in a product. Award credit for reference to any of the following points:

- Surface mount technology (SMT) allows the PCB to be manufactured by a robotic pick-and-place machine.
- Pick-and-place assembly is fast, so PCBs can be manufactured quickly.
- Pick-and-place assembly is accurate and reliable.
- Soldering will be done in a reflow oven, producing high-quality reliable joints.
- Miniature components allow high component density so the product can be made very small.
- The PCB will probably be double-sided (tracks on both sides) so complex circuits can be designed without tracks crossing.
- CAD will have been used to design and model the PCB, and this links with the CAM machines used to manufacture the PCB.

Allow any other valid point.

Sample question

Fashion and textile designers use a variety of processes and techniques to improve the structural integrity of products.

(a) Give one detailed reason why interfacing would be used on a shaped sweetheart neckline like the one shown below. [2]

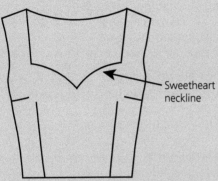

Sweetheart neckline

(b) Evaluate the use of piping and boning as methods of improving the structural integrity of textile products. [6]

ONLINE

Candidate response

(a) As the sweetheart neckline is shaped, the interfacing will stabilise fabric and prevent the fabric edge from distorting out of shape.

(b) Most fabrics are flexible and have good draping qualities; however, this often means it is difficult to achieve a good shape and structure in some textile products. Piping is a corded component that can be inserted into the seams of textile products. This adds strength to seams and helps stabilise the seam. The cord used in the piping reinforces the shape of the seam.

Boning is inserted into a casing which has been stitched onto fabric wherever support is needed in the intended product. Boning is rigid and will therefore maintain its shape during use. Corsetry, where a rigid supportive shape is needed, is a typical example of where boning is used.

Piping stabilises and strengthens seams whereas boning creates a more rigid structure. However, both improve the structural integrity of textile products.

The purpose of the interfacing has been clearly stated (to stabilise fabric) and the reason for its use (to prevent distortion) included – a detailed response worthy of the 2 available marks.

In responses to 'Evaluate' questions, evidence of appraisal or making a judgement needs to be clear. There is some evidence of this within this response. This answer would gain full marks within band 2.

The candidate demonstrates good knowledge of both piping and boning when used to improve the structural integrity of textile products and clearly states a reason why these methods would be needed.

Reference to piping is recall of facts and does not explain how piping strengthens and reinforces a seam. Further explanation is needed – for example, the extra fabric used in piping and additional stitching increase the seam's strength. The cord in the piping is often quite thick and would prevent an edge seam from folding over – this helps maintain a shape.

Boning is more clearly explained and includes reference to its practical use in products. The candidate could have included a similar example for piping – for example, to give more rigidity to the sides of a tote bag.

Mark scheme

Band 3	A coherent answer demonstrating detailed, relevant knowledge and understanding, to evaluate how designers can use piping and boning to improve the structural integrity of textile products. There will be evidence of relevant examples and well-developed, substantiated judgements in a response which is logically structured.	5–6
Band 2	The answer has some coherence, demonstrating partial knowledge and understanding, to evaluate how designers use piping and boning to improve the structural integrity of textile products. There will be some evidence of mostly relevant examples and partly substantiated judgements in a response which is generally well structured.	3–4
Band 1	The answer demonstrates only basic knowledge and understanding to evaluate how designers use piping and boning to improve the structural integrity of textile products. There will be limited evidence of relevant examples or judgements in a response which demonstrates little structure.	1–2
	Award 0 marks for incorrect or irrelevant answers.	

Product design

Sample question

Study the picture of the spice rack. The spice rack is made from beech.

It is important that designers consider the world we live in and the needs of future generations. Evaluate how designers can lessen the impact on our environment when designing and making timber-based products such as the spice rack. [6]

Candidate response

Designers should make the design as compact as possible, therefore using the least amount of material. They should make sure that the chosen timber is readily available and that it comes from a managed and sustainable source.

Designers should think about how the spice rack is to be manufactured and aim to reduce waste. They should aim to use renewable energy sources during production.

Designers should consider making the spice rack as a 'flat pack' product. This will reduce the negative environmental impact of packaging and transportation.

Designers should consider the type of finish to be applied to the spice rack. It should be durable to ensure the spice rack lasts as long as possible and also be water based so it does not harm the environment when cleaning with brushes/spray guns.

The spice rack should be clearly labelled with recycling information to encourage its environmentally safe disposal at the end of its life.

The candidate has produced a coherent answer demonstrating relevant knowledge and understanding of how designers can lessen the impact on the environment when designing and making natural timber-based products such as the spice rack.

They have included details regarding the reduction in the amount of material used and details of how to source materials from sustainable sources. The candidate has commented on the method of manufacture and the need to use power from a renewable source. They have suggested the use of the 'flat pack' method of assembly to minimise packaging and to assist in transportation. The candidate has discussed the use of environmentally friendly finishes and the need to encourage recycling.

Mark scheme

Band	Descriptor
5–6	A coherent answer demonstrating detailed, relevant knowledge and understanding to evaluate how designers can lessen the impact on the environment when designing and making timber-based products such as the spice rack. There will be evidence of relevant examples and well-developed, substantiated judgements.
3–4	An answer with some structure, demonstrating partial knowledge and understanding to evaluate how designers can lessen the impact on the environment when designing and making timber-based products such as the spice rack. There will be some evidence of relevant examples and partly substantiated judgements.
1–2	An answer demonstrating only basic knowledge and understanding to evaluate how designers can lessen the impact on the environment when designing and making timber-based products such as the spice rack. There will be limited evidence of relevant examples or judgements.
0	No response.

Glossary

A/B testing: user testing to choose between two different design ideas.

additive manufacture: computer-controlled manufacture of a 3D object by adding together materials layer by layer.

algorithm: a logical computer-based procedure for solving a problem.

amplifier: a subsystem to increase the amplitude of an analogue signal.

analogue sensor: a sensor to measure how big a physical quantity is.

annealing: a heat treatment for metal that makes it as soft as possible and reduces cracking when bending metal.

anthropometrics: the study of the sizes of people in relation to products.

aramid fibre: a non-flammable heat-resistant fibre at least 60 times stronger than nylon.

assembly line: a line of equipment/machinery manned by workers who gradually assemble a product as it passes along the line.

automated production: the use of computer-controlled equipment or machinery in manufacturing.

automation: the use of automatic equipment in manufacturing.

back emf: a high-voltage spike produced when motors, solenoids or relays are used.

batch: a limited number in a set timescale.

batch production: a number of identical or similar products are produced.

bauxite: ore containing aluminium.

bespoke: made to measure for an individual client.

bevel gears: a system to transfer the direction of rotation through 90°.

biodegradable: a material that will decompose into the Earth.

biomimicry: taking ideas from and mimicking nature.

blow moulding: a method of shaping a thermoforming polymer by blowing into a dome.

breadboard: a temporary method of constructing electronic circuits for testing.

CAD: computer-aided design.

CAM: computer-aided manufacture.

cam: a component used with a follower to convert rotary motion to reciprocating motion.

carbon neutral: no net release of carbon dioxide into the atmosphere – carbon is offset.

cellulosic fibres: natural fibres from plant-based sources.

circular economy: extracting the maximum value from resources which are then kept in use as long as possible, recovered and regenerated into new products instead of thrown away.

client: the person the designer is working for (this may or may not be the user).

closed specification: list of criteria stating what must be achieved and how it must be met.

cloud-based technology: technology that allows designers to share content via the internet.

coating: an additional outer layer added to a product.

collaboration: a number of designers working together on specific design projects.

compensation: payment given to someone as a result of loss.

compound gear train: more than one stage of gear train working together to achieve a high velocity ratio.

conductive: the ability to transmit heat or electricity.

contemporary: fashion and styles that are currently popular.

context: the settings or surroundings in which the final product will be used.

continuous flow production: identical products are being constantly made due to the high demand.

conversion: the process of cutting up a log into planks.

counterfeit: an imitation of something, sold with the intent to defraud.

crimp: the waviness in a fibre.

criteria: specific goals that a product must achieve in order to be successful.

cross filing: a method of shaping metal using files.

cross grain: runs horizontally across the fabric in line with the weft.

cultural awareness: understanding the differences in attitudes and values between people from other countries or other backgrounds.

culture: the ideas, customs and social behaviours of a particular people or society.

current: a measure of the actual electricity flowing, in amps (A).

deforestation: the removal of trees from an area of land which are not replanted.

deforming: changing the shape of a material by applying force, heat or moisture.

design fixation: when a designer limits their creativity by only exploring one avenue of design or relying too heavily on features of existing designs.

development: the creative process of selecting ideas, elements, materials and manufacturing techniques from initial ideas and using them in new ways to explore and produce newer and better designs or ideas.

digital sensor: a sensor to detect a yes/no or an on/off situation.

draw filing: a method of smoothing the edges of metal.

driven gear: the output gear from a gear train.

driver: a subsystem used to boost a signal so that it can operate an output device.

driver gear: the input gear on a gear train.

ductility: the property of a material to be able to be permanently stretched out without cracking.

eco-design: designing sustainable products that will not harm the environment by considering the effects of the technology, processes and materials used.

effort: the input force on a lever.

electroluminescent: materials that provide light when exposed to a current.

embellishments: surface decoration such as embroidery and beading.

ergonomics: the relationship between people and the products they use.

exploded drawing: a 3D drawing method to show how parts in an assembly fit together.

exploitation: the act of treating someone unfairly in order to benefit from their work.

extrusion: a length of polymer with a consistent cross-section.

fabric construction: the way a fabric has been made.

fabric specification: sets out the requirements of the fabrics needed for a product.

fad product: a product that is highly popular for only a very limited amount of time.

feedback: achieving precise control by feeding information from an output back into the input of a control system.

fibre: a fine hair-like structure.

filament: a very fine and slender thread.

finishes: added to fabrics to improve their aesthetics, comfort or function. These finishes can be applied mechanically, chemically or biologically.

finite fossil fuels: a limited amount of resources that cannot be replaced.

flowchart: a graphical representation of a program.

fluidised bath: blowing air through a powder to cause it to behave like a fluid.

focus group: group of people used to check a product design is on track.

force: a push, a pull or a twist.

Forest Stewardship Council (FSC): organisation that promotes environmentally appropriate, socially beneficial and economically viable management of the world's forests.

frequency: the number of pulses produced per second, in hertz (Hz).

fulcrum: the pivot point on a lever.

generative design: a computer-based iterative design process that generates a number of possibilities that meet certain constraints, including potential designs that would not previously have been thought of.

geotextiles: textiles associated with soil, construction and drainage.

green timber: timber that has just been felled and contains a lot of moisture.

gsm: grams per square metre – used to measure the weight of paper.

haematite: ore containing iron.

handle: how a fabric feels when handled.

hardening: a method of heat-treating metal that makes it hard but brittle.

hardwoods: timber that comes from deciduous trees and is generally harder than softwood.

high-fidelity prototype: a detailed and very accurate prototype similar to the final product.

hydrophilic membrane: a solid structure that stops water passing through but at the same time can absorb and diffuse fine water vapour molecules.

integrated circuit (IC): a miniaturised, highly complex circuit in a single component.

interactive textiles: fabrics that contain devices or circuits that respond and react with the user.

isometric drawing: a 3D drawing method, using 30° angles for projections of depth.

iterative design: a repeated cycle of quickly implementing designs or prototypes, gathering feedback and refining the design.

jig: mechanical aid used to manufacture products more efficiently.

lay plan: how templates are laid out on fabric.

LDR: light-dependent resistor. An analogue component to sense light level.

lever: a rigid bar that pivots on a fulcrum.

life cycle: the stages a product goes through from beginning (extraction of raw materials) to end (disposal).

linear economy: raw materials are used to make a product; waste is thrown away.

linear motion: movement in a straight line.

linkage: a component to direct forces and movement to where they are needed.

load: the output force from a lever.

low-fidelity prototype: a quick prototype that gives a basic idea of a product's look or functions.

lustre: a gentle shine or soft glow.

managed forest: a forest where new trees are planted whenever one is cut down.

manufactured board: sheet of timber that has been manufactured to give certain properties.

market pull: a new product is developed in response to a demand in the market or users.

mass: very large numbers made continuously over long periods of time.

mass produced: hundreds or thousands of identical products manufactured on a production line.

mass production: large quantities of identical products are produced.

mechanical advantage: the factor by which a mechanical system increases the force.

mechanism: a series of parts that work together to control forces and motion.

microcontroller: a miniaturised computer, programmed to perform a specific task and embedded in a product.

micro-encapsulation: microscopic droplets containing various substances applied to fibres, yarns and materials including paper and card.

microfibre: an extremely fine specially engineered fibre about 100 times thinner than a human hair.

micron: one thousandth of a millimetre (0.001mm) – used to specify the thickness of card.

milling: cutting grooves and slots into metal.

modelling: trying out and testing ideas or parts of designs by making scale models.

monomer: a molecule that can be bonded to others to form a polymer.

motion: when an object moves its position over time.

natural polymers: polymers that are sourced from plants.

oblique drawing: a basic 3D drawing method, using lines projected at 45°.

obsolescence: when a product is out of date or no longer usable.

one-off: a single product.

one-off production: process used when making a prototype product.

opacity: lacking transparency or translucence.

open specification: list of criteria the product must meet but not specifying how it must be achieved.

operational amplifier (op-amp): a specialised IC amplifier component.

ore: rock which contains metal.

oscillating motion: movement back and forth in a circular path.

PAR: planed all round.

PBS: planed both sides.

perspective drawing: a realistic 3D drawing method.

phase-changing materials: droplets encapsulated on fibres and materials that change between liquid and solid within a temperature range.

piece dying: an entire length of fabric is dyed.

polymerisation: chemical reaction that causes many small molecules to join together and form a larger molecule.

preservative: a chemical treatment applied to wood to prevent biological decay.

press forming: a method of shaping a thermoforming polymer by heating a pressing into a mould.

primary data: research collected 'first hand' by yourself.

primary user: the main user of the product.

primer: the base coat of paint applied straight to the material surface.

printed circuit board (PCB): a board with a pattern of copper tracks which complete the required circuit when components are soldered on.

program: a set of instructions that tells the microcontroller what to do.

programmable interface controller (PIC): a microcontroller IC used in many products.

protein fibres: natural fibres from animal-based sources.

prototype: an early model of a product or part of a product to see how something will look or function.

PSE: planed square edge.

pulp: raw material from trees used to make paper.

qualitative data: observations and opinions about a product.

quantitative data: specific measurable data given in numerical form.

quantum tunnelling composites: materials that can change from conductors to insulators when under pressure.

rating: the maximum specified quantity a component is designed to handle.

raw edges: fabric edges that are not neatened – unfinished.

ream: pack of 500 sheets.

reciprocating motion: movement back and forth in a straight line.

reinforcement: extra material added to increase strength.

rendering: the method of using colour and shading to represent the nature of a surface.

resist method: a means of preventing dye or paint from penetrating an area on the fabric. This creates the patterns.

rigid: inflexible, stiff.

rotary motion: movement in a circular path.

rotational velocity: the number of revolutions per minute (rpm) or per second (rps).

schematic diagram: a circuit diagram showing the connections between individual components.

seam allowance: the distance between the raw edge of the fabric and the stitching line.

seasoning: the process of removing moisture from newly converted planks.

secondary data: research collected by others.

self-finished: a material that does not require the application of a finish to protect it or improve its appearance.

selvedge: the sealed edge of the fabric.

shot-blasting: using grit, fired at high pressure, to clean a surface by abrasion.

simple gear train: two spur gears meshed together.

smelting: the process of extracting metal from ore.

softwood: timber that comes from coniferous trees and is generally less expensive than hardwoods.

spur gear: a gear wheel with teeth around its edge.

stakeholder: a person other than the main user who comes into contact with or has an interest in the product.

straight grain: indicates the strength of the fabric in line with warp yarns.

subroutine: a small subprogram within a larger program.

subsystem: the interconnected parts of a system.

surface decoration techniques: used to improve the aesthetics of a product by adding colour, texture and pattern, for example dyeing, printing and embroidery.

surface mount technology (SMT): the industrial method of mounting miniature components onto a PCB using robotic machines.

sustainability: meeting today's needs without compromising the needs of future generations.

synthetic: derived from petrochemicals or manmade.

synthetic polymers: polymers that are sourced from crude oil.

system: a set of parts which work together to provide functionality to a product.

system diagram: shows the interconnections between subsystems in an electronic system.

systems thinking: considering a design problem as a whole experience for the user.

technology push: products developed as a result of new technology.

tempering: a method of heat-treating metal that reduces brittleness.

tessellate: fit together pattern templates to use the least amount of fabric.

tessellation: an arrangement of shapes closely fitted together in a repeated pattern without gaps or overlapping.

thermistor: an analogue component to sense temperature.

thermoforming polymer: a polymer that can be reheated and reformed.

thermosetting polymer: a polymer that cannot be reformed with heat.

throwaway society: a society that excessively consumes and wastes resources.

tolerance: an allowance included in the seam allowance for inconsistency when assembling a product.

torque: a turning force.

turning: a method of producing cylinders and cones using a centre lathe.

ultraviolet (UV) light: outside the human visible spectrum at its violet end.

user: the person or group of people a product is designed for.

user-centred design (USD): looking at and checking the needs, wants and requirements of the user at every stage of the design process.

user testing: testing by observing a user interact with and use your product for its intended purpose.

vacuum forming: a method of shaping a thermoforming polymer sheet by heating around a former.

velocity ratio: the factor by which a mechanical system reduces the rotational velocity.

veneer: thin sheet of natural timber.

voltage: the electrical 'pressure' at a point in a circuit, in volts (V).

voltage gain: the amplification factor of an amplifier subsystem.

warp: yarns that run along the length of the fabric.

wastage: the process of shaping material by cutting away waste material.

weave: the pattern woven in the production of fabric.

weft: yarns that run across the fabric.

worm drive: a compact gear system which achieves a very high velocity ratio.

yarn: spun thread used for knitting, weaving or sewing.

Acknowledgements

Picture credits

The Publishers would like to thank the following for permission to reproduce copyright material.
Fig.1.1 © Stepan Popov/123 RF; Fig.1.3 © Lucadp/stock.adobe.com; Fig.1.4 © Ulldellebre/stock.adobe.com; Fig.1.5 © Dee Cercone/Everett Collection/Alamy Stock Photo; Fig.1.6 © Simon Belcher/Alamy Stock Photo; p.13 t © petovarga/123 RF; b © petovarga/123 RF; Fig.1.7 © Plus69/stock.adobe.com; Fig.1.9 © Alexandr Bognat/stock.adobe.com; Fig.1.10 © dpa picture alliance/Alamy Stock Photo; Fig.1.11 © Andreas von Einsiedel/Alamy Stock Photo; Fig.1.12 © Dyson; Fig.1.13 Ruby Tree Collection designed by Bethan Gray; Fig.2.8 Dan Hughes; Fig.2.13 © Nikkytok/stock.adobe.com; Fig.2.14 © Vladimir/stock.adobe.com; Fig.2.22 © Andrey Popov/stock.adobe.com; Fig.2.25 © Raymond McLean /123RF; Fig.3.7 © Sergofoto/123RF; Fig.3.8 © Chamillew/stock.adobe.com; Fig.3.10 © Colin Moore/123 RF; Fig.3.11 © RichLegg/E+/Getty Images; Fig.3.13 © Andreja Donko/stock.adobe.com; Fig.3.15 © Anton Oparin/123RF; Fig.3.16 © Nataliia Pyzhova/stock.adobe.com; Fig.3.17 © Cherryandbees/stock.adobe.com; Fig.3.18 Jacqui Howells; Fig.3.19 © Pincasso/stock.adobe.com; Fig.3.23 Jacqui Howells; Fig.3.24 Jacqui Howells; Fig.3.25 © Tapui/stock.adobe.com; Fig.3.26 Jacqui Howells; Fig.3.29 © vvoe/stock.adobe.com; Fig.3.30 © Claudette/Stockimo/Alamy Stock Photo; Fig.4.4 © Wichien Tepsuttinun/Shutterstock.com; Fig.4.8 © Unkas Photo/Shutterstock.com; Fig.4.9 © Anton Starikov/Shutterstock.com; Fig.4.14–Fig. 4.19 Ian Fawcett; Fig.4.20 © Treboreckscher/123 RF; Fig.4.22 © C R CLARKE & CO (UK) LIMITED; Fig.4.23 © Oleg/stock.adobe.com; Fig.4.32 © Alexlmx/stock.adobe.com; Fig.4.35 © Hoda Bogdan/stock.adobe.com; Fig.4.42 © Andrey Eremin/123RF; Fig.4.43 © Shaffandi/123 RF; Fig.4.44 © Barylo Serhii/123RF; Fig.4.45 © Bravissimos/stock.adobe.com; Fig.4.46 © Sorapol Ujjin/123RF; Fig. 4.47 © Okinawakasawa/stock.adobe.com; Fig.4.53 © Showcake/stock.adobe.com; Fig.4.54 © © gl0ck33/123 RF; Fig.4.55 © Paul Broadbent/Alamy Stock Photo; Fig.4.56 © Belchonock/123RF; Fig.4.57 © Ajay Shrivastava/stock.adobe.com; Fig.4.58 Ian Fawcett; Fig.4.59 Ian Fawcett; p.121 © Aleksandr Rado/123RF; Fig.5.1 © Wolf Safety Lamp Company; Fig.5.2 © Pixtour/stock.adobe.com; Fig.5.9 Dan Hughes; Fig.5.10 Andy Knight; p.139 © KITEMARK and the Kitemark device are reproduced with kind permission of The British Standards Institution. They are registered trademarks in the United Kingdom and in certain other countries; p.141 l © pixelrobot/stock.adobe.com; r © Rafa Irusta/stock.adobe.com; p.142 l Chris Walker; r © Arvind Balaraman/123 RF; p.144 © Michał Dzierżyński/123 RF.